Sol-Gel Commercialization and Applications

Related titles published by The American Ceramic Society:

Boing-Boing the Bionic Cat and the Jewel Thief
By Larry L. Hench
©2001, ISBN 1-57498-129-3

The Magic of Ceramics
By David W. Richerson
©2000, ISBN 1-57498-050-5

Boing-Boing the Bionic Cat
By Larry L. Hench
©2000, ISBN 1-57498-109-9

*Innovative Processing and Synthesis of Ceramics, Glasses, and Composites IV
(Ceramic Transactions Volume 115)*
Edited by Narottam P. Bansal and J.P. Singh
©2000, ISBN 1-57498-111-0

*Innovative Processing and Synthesis of Ceramics, Glasses, and Composites III
(Ceramic Transactions Volume 108)*
Edited by J.P. Singh, Narottam P. Bansal, and Koichi Niihara
©2000, ISBN 1-57498-095-5

Advances in Ceramic Matrix Composites V (Ceramic Transactions, Volume 103)
Edited by Narottam P. Bansal, J.P. Singh, and Ersan Ustundag
©2000, ISBN 1-57498-089-0

Ceramic Innovations in the 20th Century
Edited by John B. Wachtman Jr.
©1999, ISBN 1-57498-093-9

Advances in Ceramic Matrix Composites IV (Ceramic Transactions, Volume 96)
Edited by J.P. Singh and Narottam P. Bansal
©1999, 1-57498-059-9

*Innovative Processing and Synthesis of Ceramics, Glasses, and Composites II
(Ceramic Transactions, Volume 94)*
Edited by Narottam P. Bansal and J.P. Singh
©1999, ISBN 1-57498-060-2

*Innovative Processing and Synthesis of Ceramics, Glasses, and Composites
(Ceramic Transactions, Volume 85)*
Edited by Narottam P. Bansal, Kathryn V. Logan, and J.P. Singh
©1998, ISBN 1-57498-030-0

Advances in Ceramic Matrix Composites III (Ceramic Transactions, Volume 74)
Edited by Narottam P. Bansal and J.P. Singh
©1996, ISBN 1-57498-020-3

For information on ordering titles published by The American Ceramic Society, or to request a publications catalog, please contact our Customer Service Department at 614-794-5890 (phone), 614-794-5892 (fax),<customersrvc@acers.org> (e-mail), or write to Customer Service Department, 735 Ceramic Place, Westerville, OH 43081, USA.

Visit our on-line book catalog at <www.ceramics.org>.

Ceramic Transactions
Volume 123

Sol-Gel Commercialization and Applications

Proceedings of the Symposium on Sol-Gel Commercialization and Applications at the 102nd Annual Meeting of The American Ceramic Society, held May 1–2, 2000, in St. Louis, Missouri.

Edited by
Xiangdong (Shawn) Feng
Ferro Corporation

Lisa C. Klein
Rutgers University

Edward J.A. Pope
MATECH Advanced Materials

Sridhar Komarneni
The Pennsylvania State University

Published by
The American Ceramic Society
735 Ceramic Place
Westerville, Ohio 43081
www.ceramics.org

Proceedings of the Symposium on Sol-Gel Commercialization and Applications at the 102nd Annual Meeting of The American Ceramic Society, held May 1–2, 2000, in St. Louis, Missouri.

Copyright 2001, The American Ceramic Society. All rights reserved.

Statements of fact and opinion are the responsibility of the authors alone and do not imply an opinion on the part of the officers, staff, or members of The American Ceramic Society. The American Ceramic Society assumes no responsibility for the statements and opinions advanced by the contributors to its publications or by the speakers at its programs. Registered names and trademarks, etc., used in this publication, even without specific indication thereof, are not to be considered unprotected by the law.

No part of this book may be reproduced, stored in a retrieval system, or transmitted in any form or by any means, electronic, mechanical, photocopying, microfilming, recording, or otherwise, without written permission from the publisher.

Authorization to photocopy for internal or personal use beyond the limits of Sections 107 and 108 of the U.S. Copyright Law is granted by the American Ceramic Society, provided that the appropriate fee is paid directly to the Copyright Clearance Center, Inc., 222 Rosewood Drive, Danvers, MA 01923 USA, www.copyright.com. Prior to photocopying items for educational classroom use, please contact Copyright Clearance Center, Inc.

This consent does not extend to copying items for general distribution or for advertising or promotional purposes or to republishing items in whole or in part in any work in any format.

Please direct republication or special copying permission requests to Copyright Clearance Center, Inc., 222 Rosewood Drive, Danvers, MA 01923 USA 978-750-8400; www.copyright.com.

Cover photo: "SEM micrograph of $Sr_{0.48}Ba_{0.52}Nb_2O_6$ gel fiber" is courtesy of Masahiro Toyoda and Kyougo Shirono, and appears as figure 2 in their paper "Preparation and Characterization of $Sr_{0.48}Ba_{0.52}Nb_2O_6$ Ceramics Fibers Through Sol-Gel Processing," which begins on page 111.

Library of Congress Cataloging-in-Publication Data
A CIP record for this book is available from the Library of Congress.

For information on ordering titles published by The American Ceramic Society, or to request a publications catalog, please call 614-794-5890.

4 3 2 1–04 03 02 01

ISSN 1042-1122
ISBN 1-57498-120-X

Contents

Preface vii

The Coming Golden Age of Sol-Gel Technology: Maximizing
Value in the Age of Biotechnology and the Internet 1
   Edward J.A. Pope

Functional Coatings as an Interesting Tool for Industrial
Opportunities: Development and Commercialization 15
   Helmut Schmidt

Sol-Gel Commercialization in Japan 29
   Sumio Sakka

Molecular Templated Sol-Gel Synthesis of
Nanoporous Dielectric Films 39
   Suresh Baskaran, Jun Liu, Xiaohong Li, Glen E. Fryxell, Nathan Kohler,
   Christopher A. Coyle, Jerome C. Birnbaum, and Glen Dunham

Titanium Oxo-Organo Clusters: Precursors for the Preparation
of Nanostructured Titanium Oxide Based Materials 49
   N. Steunou, C. Sanchez, P. Florian, S. Förster, C. Göltner, and M. Antonietti

Proton Conducting SiO_2-P_2O_5-ZrO_2 Sol-Gel Glasses 73
   M. Aparicio and L.C. Klein

The Study of Hydrolysis and Condensation of Mixed
Solution from $Si(OC_2H_5)_4$ and $Zr(O-nC_3H_7)_4$ 81
   Dae-Yong Shin, Sang-Mok Han, Kyung-Nam Kim, Wi-Soo Kang,
   and Sang-Kyu Kang

Patterned Microstructure of Sol-Gel Derived Complex
Oxides Using Soft Lithography 87
    Seana Seraji, Yun Wu, Nels Jewell-Larson, Mike Forbess,
    Steven Limmer, and Guozhong Cao

$SrBi_2(Nb,V)_2O_9$ Ferroelectric Films by Sol-Gel Processing 93
    Y. Wu, S. Seraji, M.J. Forbess, S.J. Limmer, and G.Z. Cao

Flexible Sheets of Dimethysiloxane-Based
Inorganic/Organic Hybrids 99
    Shingo Katayama, Keiko Kawakami, and Noriko Yamada

Achievement of Crack-Free Ceramic Coatings over
1 μm in Thickness via Single-Step Deposition 105
    Hiromitsu Kozuka, Katsumi Katayama, Yoshiro Isota,
    and Shinsuke Takenaka

Preparation and Characterization of $Sr_{0.48}Ba_{0.52}Nb_2O_6$
Ceramic Fibers Through Sol-Gel Processing 111
    Masahiro Toyoda and Kyougo Shirono

Sol-Gel Processing for the Lanthanides 117
    James L. Woodhead

Sol-Gel Derived Nickel Titanate for Tribological Coatings 121
    D.J. Taylor, P.F. Fleig, and R.A. Page

Synthesis of Wood-(SiO_2,Al_2O_3) Inorganic
Composites by Sol-Gel Process 127
    Xicheng Wang, Shulan Shi, Hongmei Hou, Peng Shi, and Li Sun

Real-Time Monitoring of Striation Development
during Spin-on-Glass Deposition 133
    Dylan E. Haas and Dunbar P. Birnie III

Index 139

Preface

Sol-gel, a low-temperature process used to chemically produce glasses, ceramics, and composites, has been touted as a new, exciting, and potentially useful technique for the preparation of high-performance materials. The explosive growth of sol-gel science is demonstrated by the exponential increase in the number of papers published on the topic: 40 between 1970–1980, 300 by 1984, 6000 by 1994, and 59,000 by 1999. Many industrial companies now have programs in sol-gel research and its development. Yet there are still only a handful of successful sol-gel commercial products.

This symposium on Sol-Gel Commercialization and Applications was held in St. Louis, Missouri, May 1–2, 2000, in conjunction with the 102nd Annual Meeting of The American Ceramic Society. This book, Sol-Gel Commercialization and Applications, is the proceedings of that symposium. With 42 presentations and more than 100 in attendance, this symposium featured sessions on sol-gel commercial experiences and issues, sol-gel for electrical and display applications, and sol-gel synthesis of powders, ceramics, coatings, and catalysts.

We, the editors, would like to thank the session organizers and session chairs, Helmut Schmidt, Jun Liu, and John P. Cronin, for making the symposium a success. Our appreciation also goes to Ms. Marie Wentz for secretarial support in publishing this book.

Xiangdong (Shawn) Feng
Lisa C. Klein
Edward J.A. Pope
Sridhar Komarneni

THE COMING GOLDEN AGE OF SOL-GEL TECHNOLOGY:
Maximizing Value in the Age of Biotechnology and the Internet.

Edward J. A. Pope, Ph.D.
MATECH Advanced Materials
31304 Via Colinas, Suite 102
Westlake Village, CA 91362

ABSTRACT

Recent market studies estimate that sol-gel technology is a $400 million per year technology sector that is rapidly growing. The vast majority of the current applications of sol-gel technology, however, are in relatively low tech, low value-added industrial materials and applications. These include nanophase ceramic powders and other raw materials, industrial coatings, structural glass and ceramic fibers, abrasives, and porous catalyst supports, all of which are highly price sensitive applications. Yet untapped are high value added applications, such as biological sensing and medical diagnostics, adaptive optics, optical computing and integrated optics devices, flat panel displays, biotech drug delivery, biomedical implants, and bioartificial organs (cell therapy). These are some of the emerging twenty-first century applications that will dwarf the existing sol-gel markets.

INTRODUCTION

What defines a "Golden Age" for a technology? There are many possible criteria. Typically, a golden age has a dramatic impact on elevating and advancing the human condition. Great technological progress and innovation often occurs. Often, the new age stimulates a period of entrepreneurship, industry building, and risk taking. As a consequence, both wealth creation and job creation occurs. In this century, lots of government and private research and development dollars are also funneled into new "breakthrough" technology sectors [1].

Historically, Golden Ages have been *based on new materials!* The ancient ages have included; the Stone Age, the Bronze Age, and the Iron Age. Then came the Industrial Revolution (1700s-1900s), the ultimate extension and fulfillment of the iron age. In this century, we've seen the Space Age (1950s - '70s), the Computer Age (1960s - '90s), the Information Age (late-1990's), and now the Biotech Age (2000 – 2020?). All of these recent "ages" have been enabled by new materials and materials processing technologies.

To the extent authorized under the laws of the United States of America, all copyright interests in this publication are the property of The American Ceramic Society. Any duplication, reproduction, or republication of this publication or any part thereof, without the express written consent of The American Ceramic Society or fee paid to the Copyright Clearance Center, is prohibited.

THE ECONOMIC FOOD CHAIN

So, if materials science is so important, why aren't materials scientists as rich as Bill Gates? Simple, while materials science may be at the core of ushering in new advances in technology, materials themselves are quite low on the economic food chain (see Figure 1).

Figure 1: The economic food chain.

Most materials companies compete on price. Often, there are a wide number of alternate choices for high quality materials that can be fabricated into finished components. Consider, for example, the personal computer (PC). The most important single component is the CPU, which typically costs a few hundred dollars, depending upon how recent and advanced it is. The basic raw materials are a single crystal silicon substrate, which can be obtained from multiple vendors, mostly in Southeast Asia, and a ceramic substrate, usually alumina. The alumina substrates cost about 25 cents each or less (see Figure 2). Because there are numerous vendors, wafers are also relatively inexpensive. The bulk of the cost of the CPU is the result of the value added that is provided by the circuit architecture and performance characteristics. Hence, companies like INTEL and AMD are quite profitable and enjoy very high market capitalizations, so long as they keep advancing their technologies forward. The PCs themselves are manufactured by over a dozen major brands and, hence, are highly price sensitive and low margin commodity products.

The take home lessons from the preceding can be summarized as: 1) Focus upon high value added applications & end products; 2) Avoid the "raw materials" business; 3) Integrate unique material technologies with "the product"; 4) Position your business model as close to the market as possible, and; 5) NEVER COMPETE ON COST!!!!!

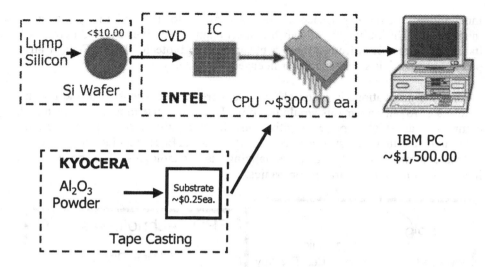

Figure 2: The economic food chain as applied to PC manufacturing.

IMPLICATIONS FOR SOL-GEL PROCESSING

Sol-gel technology is not a specific material but a process. Bearing in mind the concept of where the intended application of the technology resides in the economic food chain, several guidelines can be postulated. First, avoid applications for sol-gel that can already be performed well by other technologies. Unless you're Corning or 3M, incremental reductions in cost are insufficient to justify the effort. Only companies with deep pockets and large economies of scale can afford to play that game. Innovators (both academic and industrial) should focus on exploiting what sol-gel does "naturally" with little effort. Moreover, they should concentrate on integrating sol-gel with other "high value added" disciplines, such as advanced optics, biotechnology, and medicine.

MATECH'S STRATEGY

MATECH, which was founded in 1989 primarily as a contract research and development laboratory, has conducted fundamental research into sol-gel derived materials. In addition to contract research, MATECH has developed a number of innovative processes and materials as part of its IR&D programs. Initially, MATECH sought to commercialize these technologies through out-licensing to much larger corporations. After a few years, it became apparent that, with few exceptions, large corporations are not interested in "early stage" research innovations. This experience is not dissimilar to that of most academic institutions. The University of California, for example, derives over 95 percent of its licensing revenues from a mere 5 patents, all in biotechnology. Few, if any, patents in materials science are ever licensed.

Even if innovations in materials science were easy to license out, the returns in terms of royalty payments would be slim. This is because most materials based companies are highly price sensitive and have relatively low profit margins compared to other industry sectors. In 1997, MATECH decided to pursue an alternate strategy towards commercializing its technologies – to become an incubator of high tech start-up companies.

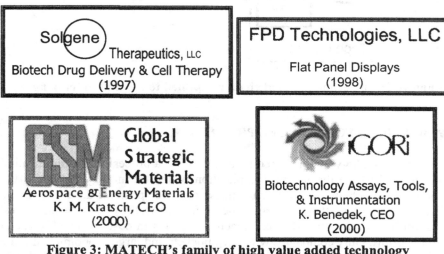

Figure 3: MATECH's family of high value added technology "spin-off" companies utilizing sol-gel technology.

The first company launched was Solgene Therapeutics in late-1997, aimed at biotech drug delivery and cell therapy. Biotech drug sales exceed $25 billion per year worldwide. Moreover, biotech drug sales are increasing by about 15 percent per year. In excess of 400 biotech drugs are currently in FDA clinical trials. With the successful completion of the Human Genome Project, many hundreds, if not thousands, of new biotech therapeutics will be discovered. They all need to be delivered. Moreover, the majority cannot be delivered effectively by conventional drug delivery technologies. Enter Solgene with its unique sol-gel derived cell encapsulation technology. Instead of producing drugs in a factory, like the way AMGEN produces erythropoietin (aka EPOGEN^{TM}), the "biotech factory" can be implanted into the patient to manufacture the drug in real time continuously for many months. This obviates the necessity for frequent (usually daily) injections.

Another of MATECH's spin-offs is FPD Technologies, LLC, formed in 1998. FPD's mission is to develop and commercialize next generation flat panel display technologies. Anyone who owns a laptop knows how annoying the current state-of-the-art active matrix liquid crystal displays (AMLCDs) are to view and utilize. The poor angle of viewing, low brightness, and general lack of vibrancy in the colors greatly diminish their desirability. Nonetheless, flat panel displays are expected to exceed $15 billion in sales this year. FPD's objective is to make flat panel displays using fluorescence, the same way CRTs produce the image for the viewer. Using highly efficient sol-gel derived fluorescent coatings, the image produced can be as bright and vibrant as a bulky CRT, but just as compact as an AMLCD.

WHAT COMES NATURALLY TO SOL-GEL?

For over two decades, sol-gel researchers have been attempting to utilize sol-gel for the synthesis and fabrication of a wide range of glass and ceramic engineered materials, including laser glasses [2], high temperature glass-ceramic coatings [3], optical fibers and performs [4], and optical coatings [5]. In most instances, these types of application involve high temperature heat treatment and/or densification and sintering. The original gel, however, is highly porous, wet, and amorphous. Therefore, in fabricating these high temperature processed products, the inherent properties of the gel must be overcome.

In many of the new and emerging applications of sol-gel technology, some of the attributes of gels that come "naturally" can be exploited. These include: 1) Gels can be prepared at ambient temperatures; 2) Wet gels are highly porous; 3) Gels form well in aqueous solutions; 4) Both organic and inorganic molecules, dopants, and particulates can be readily dissolved or dispersed in gels; 5) Gels can be dried/processed at $T<100^{\circ}C$; 6) Dried gels are typically a high surface area, porous, amorphous solid, and; 7) Dried gels are often optically transparent.

One of the attributes of wet gels and dried aerogels is the very high porosity and surface areas that can be routinely achieved. In Figure 4, an artist's 3-Dimensional representation of silica gel is shown. It is based upon T.E.M. studies conducted on gels during the late-1980's [6]. In Table 1, the physical properties of six different xerogel and aerogel silicas are presented, illustrating the ability to tailor pore sizes across a wide size range and the exceptionally high surface areas which can also be obtained.

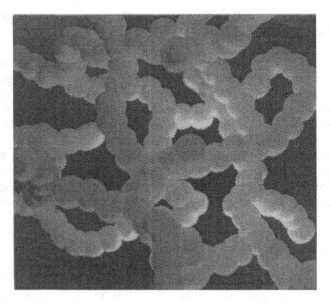

Figure 4: 3-dimensional conceptual view of silica gel structure (at ~250,000X magnification).

Table 1: Examples of six types of porous silica gels manufactured by MATECH.

MATERIAL TYPE	PORE SIZE (Angstroms)	SURFACE AREA (m^2/gm)	PORE VOLUME (cc/gm)
A	17	400	0.3
B	100	500	0.7
C	160	900	2.2-3.0
D	250	1100	2.2-3.0
E	500	450	2.0-3.0
F	1000	400	1.5-2.0

The porosity of silica gels can be exploited in many different ways, one of which is to impregnate an organic monomer within the pores and polymerize them *in situ*. When the gel phase is processed at temperatures near ambient, the opportunity exists to be able to incorporate optically active organic species, such as laser dyes, into either the inorganic phase and/or the organic phase, or both. In

Figure 5, a wide assortment of organic and inorganic fluorophor doped Silica/PMMA nanocomposites are shown under UV illumination.

Figure 5: Transparent optically active (fluorescent) dye doped and lanthanide doped silica gel/polymer nanocomposites.

SOL-GEL OPTICS: SENSORS & BIOSENSORS

The application of sol-gel to sensors and biosensors has been extensively reviewed recently [7]. Various fluorescent bioactive molecules have been incorporated in porous sol-gel microbeads which can be used to sense pH, ethanol concentration, organophosphonate concentration, heavy metal ions, antibodies, glucose, and other species [7]. These sensors can be made inexpensively and be disposable.

SOL-GEL OPTICS: WHOLE CELL BIOSENSORS

Living cells manifest a wide range of highly sensitive metabolic processes and represent an opportunity to develop highly sensitive biological sensors. Challenges to developing whole cell based sensors include keeping them alive and interfacing with the cell's metabolic functions. Nonetheless, whole cell biosensing is emerging as an exciting new area of research and development. The issue of keeping the cells alive can be mitigated in *in vivo* sensing applications. Palti has patented the use of living tissue cells as sensors for blood and constituent levels, such as glucose monitoring [8]. One drawback to *in vivo* applications is the need to immunoisolate the foreign cells to avoid immunorejection reactions. Researchers at Stanford have already demonstrated how to make simple non-immunoisolated sensors from living cells [9,10]. In their work, they demonstrated

ATP measurement and detection among other things.

The issue of immunoisolation has been largely resolved by our research into microbial and mammalian tissue cell encapsulation [11-15]. The majority of our research, which has now been spun-off into Solgene Therapeutics, LLC, has been centered around biotech drug delivery and cell therapy. For example, silica gel encapsulated pancreatic islet allografts have been successfully transplanted into severely diabetic mice, resulting in a complete remission of symptoms (glucosuria and high hematological glucose levels) for in excess of four months [11,15]. No rejection of the encapsulated foreign tissue was observed. Moreover, recent results obtained at Cornell indicate no systemic immunological response to the silica gel encapsulant (unpublished).

In our earliest work on cell encapsulation, the single cell fungi *S. cerevisiae* was stained with pyranine as a means of monitoring alcohol evolution during fermentation after encapsulation with silica gel [16,17]. In this manner, we were able to optically "interface" with the living cells by monitoring changes in the fluorescence emission spectra. Thus, for *in vivo* applications, the solution to both key challenges of keeping the cells alive and interfacing with their metabolic functions has been demonstrated.

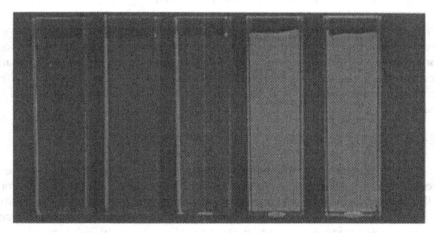

Figure 6: Genetically-altered *E. coli* with GFP (Affymax).

CELL ENCAPSULATION FOR HIGH THROUGHPUT SCREENING & NEW DRUG DISCOVERY.

Researchers at ORNL have recently demonstrated the ability to attach a genetically engineered microorganism, *Pseudomonas fluorescens* HK44, to a hybrid circuit and detect ppb levels of naphthalene [18]. Their "critter on a chip" technology, if combined with recent cell encapsulation advances, could lead to the development of *living* biosensor arrays for *in vivo* medical diagnostics.

Another major development has been the incorporation of green fluorescent protein (GFP) in genetically engineered *E. coli*, pioneered by Affymax. This genetically engineered organism allows for fluorescence to be used for bioactivity assays. Cultures which have been engineered to express various levels of GFP are shown in Figure 6.

Live cell based bioactivity assays in 96 wellplate formats are already utilized as part of quality control (QC) procedures for acceptance of biotech drugs by many major pharmaceutical and biotech companies. AMGEN, for example, utilizes live cell assays, with radionucleotide labeling, for validation of their blockbuster drugs EPOGEN™ and NEUPOGEN™. These types of assays usually involve single layers of cells plated on the bottom of each well.

Encapsulated cells in silica gel permit higher cell counts to be utilized, thereby increasing the sensitivity of these assays. Moreover, the distribution of the cells can be uniform and fixed, resulting in higher signal to noise ratios. For example, cancer cell lines in 96, 386, 1536 well microtiter plate formats can be used to conduct high throughput screening (HTS) of potential anti-cancer agents.

Encapsulated T-cell (CD4+) lines could be used for screening reverse transcriptase (RT), protease, & integrase inhibitors. Also, these assays can be used to screen for specific inducing factors for promoter gene regulated protein expression, critically important for cell therapy based biotech drug delivery technologies.

SOLGENE THERAPEUTICS: BIOTECH DRUG DELIVERY

The ultimate in high-value-added sol-gel technology could arguably be the sustained delivery of blockbuster biotech drugs. With the completion of the "first draft" of the Human Genome, announced on June 26, 2000, hundreds or perhaps thousands of new biotech compounds will be discovered in the next decade. Solgene Therapeutics' mission is to improve the efficacy, reduce the toxicity, and broaden the treatment options for *known and proven* biotech drugs. In addition, to *enable* the delivery of known drugs which have failed due to inadequate delivery. And, ultimately, to *fulfill* the promise of the Human Genome Project.

The Achilles Heal of biotech drugs is inadequate delivery. Nature Biotechnology, in a lead editorial wrote, "Drug discovery may be biotech's *raison d'etre*, but its success is utterly dependent upon versatile delivery."[19]. Biotech drug delivery technologies today are essentially "bolt-ons", that is to say, afterthoughts. The drug is manufactured by traditional biotech techniques and the delivery technology is added as the last step, usually with little or no modification of the drug.

Solgene's approach is the complete and total integration of drug production with drug delivery. By encapsulating genetically altered human cells in an immunoisolating silica gel encapsulant, these tiny biotech factories can be implanted directly into the patient, delivering therapeutic compounds for many months per single injection.

The ability to deliver varied biotech compounds without the immune system reacting to the allograft cells is due to the uniquely tailored pore structure of the gel. In sol-gel technology, we have the ability to create gels with pore sizes ranging from 1.0 nm to 10 microns. By tailoring the pore structure to be "just right", Solgene was able to develop materials that can deliver virtually all biotech drugs effectively (see Figure 7).

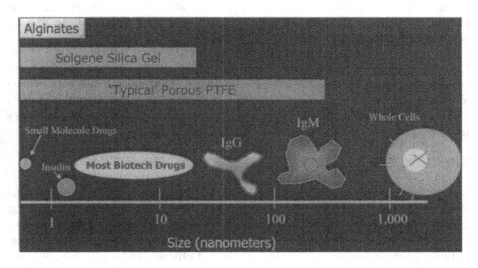

Figure 7: Comparison between Solgene's silica gel encapsulants and the two leading competitive material systems.

As described earlier, Solgene's technology has already been demonstrated for diabetes, resulting in complete euglycemia for 4 months in a severely diabetic mouse model [11,15]. Also demonstrated was the ability to successfully encapsulate genetically altered cells. In Figure 8, results from lac-z transfected mesenchymal stem cells are shown. On the left are control cell (no gene), which remain white upon exposure to an indicator dye. On the right are cells which have the gene and, hence, produce beta-galactosidase, which turns the indicator dye deep blue.

More recently, Solgene has selected erythropoietin as its "lead" drug delivery candidate, dubbed EPO-90™. To be announced shortly is a licensing agreement for EPO secreting human cell lines from a major international pharmaceutical company.

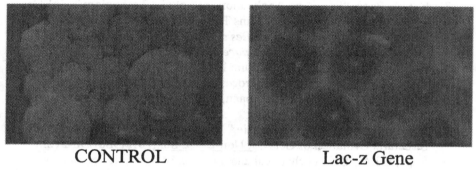

CONTROL Lac-z Gene

Figure 8: lac-z transfected mouse mesenchymal stem cells after 10 days in vitro culture (with Joseph M. Lane, M.D. & Andy Tomin).

SUMMARY

In short, it should be remembered that Sol-Gel is a *process* and not a product. How and what we apply the process to dictates the value added. In other words, the value of sol-gel depends upon the value of the product made using the process. Fortunately, the natural characteristics of sol-gel derived materials are highly compatible with biotech and medical applications which are extremely *high value added!* Opportunities for Sol-Gel encapsulation technology include Fiber-Optic Chemical and Biological Sensors, Integrated Optic Sensor Arrays, Micro-Bead based Enzyme Reactor Technology (MicroBERT™), Whole Cell Biosensors (WCB), WCB-based High Throughput Screening (e.g. cancer, HIV drugs, etc.), and Biotech Drug Delivery & Cell Therapy. Tremendous growth in these and other areas will fuel Sol-Gel's Golden Age!

REFERENCES

1. Geoffrey Moore, "Living on the Fault Line: Managing Shareholder Value in the Age of the Internet", (HarperPublishing, New York, 2000).
2. E.J.A. Pope and J.D. Mackenzie, "Sol-Gel Synthesis of Neodymia-Silica Laser Glass", Extended Abstract in <u>Structure-Property Relationships in Optical Materials</u>, ed. by D. Eimerl et al, (Materials Research Society EA-12, Pittsburg:1987)27-28.; E. J. A. Pope and J. D. Mackenzie,"Nd-Doped Silica Glass I: Structural Evolution in the Sol-Gel State", <u>J. Non-Cryst. Sol.</u>, 106(1988)236-241.; E. J. A. Pope and J. D. Mackenzie,"Sol-Gel Processing of Neodymia-Silica Glass", <u>J. Amer. Ceram. Soc.</u>, 76[5](1993)1325-28.
3. E. J. A. Pope, "Emerging Applications of Sol-Gel Technology", <u>Sol-Gel Production</u>, H. Schmidt, ed. (Trans Tech Publishing, Ltd., Switzerland, 1998). *This paper describes studies on Y-Zr-Al-Si-O high temperature glass-ceramic coatings for aerospace applications.*
4. U. S. Patent 5,023,208, issued June 11, 1991, E. J. A. Pope, Y. Sano, S. Wang, and A.Sarkar, "Sol-gel Process for Glass and Ceramic Articles"
5. Dislich, H., Hinz, P., and Kaufmann, R., FRG-Patent 19 41 181; August 13, 1969.
6. E. J. A. Pope and J. D. Mackenzie," Theoretical Modelling of the Structural Evolution of Gels", <u>J. Non-Cryst. Sol.</u>,101(1988)198-212.
7. Pope, E.J.A.,"Sol-gel chemical and biological fiber-optic sensors and sensor arrays" *Sol-Gel Optics V*, B.S.Dunn, E.J.A.Pope, M.Yamane, and H.Schmidt, eds. (SPIE Proc. Vol. 3943, 2000).
8. Y. Palti, "System for Monitoring and Controlling Blood and Tissue Constituent Levels", U. S. Patent 5,368,028 (November 29, 1994). *This patent traces itself back to a patent issued August 11, 1989, U. S. Patent 5,101,814.*
9. J. B. Shear, et al., "Single Cells as Biosensors for Chemical Separations", <u>Science</u>, 267 (6 January 1995) pp. 74-77.
10. "Making a Biosensor from a Cell and a Fluorescent Dye", <u>Biophotonics</u>, (March/April 1995) p.17.
11. E. J. A. Pope, K. Braun, and C. M. Peterson, "Bioartificial Organs I: Silica Gel Encapsulated Pancreatic Islets for the Treatment of Diabetes Mellitus", <u>J. Sol-Gel Sci. & Tech.</u>, 8 (1997) pp. 635-639.
12. E. J. A. Pope, "Encapsulation of Living Tissue Cells in an Organosilicon", U. S. Patent 5,693,513 (December 2, 1997).
13. E. J. A. Pope, "Encapsulation of Animal and Microbial Cells in an Inorganic Gel Prepared from an Organosilicon", U. S. Patent 5,739,020 (April 14, 1998).

14. E. J. A. Pope, U. S. & foreign patents pending.
15. K. P. Peterson, C. M. Peterson, and E. J. A. Pope, "Silica Sol-Gel Encapsulation of Pancreatic Islets", <u>Proc. Soc. Exp'l Bio. & Med.</u>, accepted, in press.
16. E. J. A. Pope, "Gel Encapsulated Microorganisms: *Saccharomyces cerevisiae* - Silica Gel Biocomposites", <u>J. Sol-Gel Sci. & Tech.</u>, 4 (1995) pp.225-229.
17. E. J. A. Pope,"Living Ceramics" in <u>Sol-Gel Science and Technology</u>, edited by E. J. A. Pope, S. Sakka, and L. C. Klein (ACerS Transactions Vol. 55, Westerville, OH, 1995)pp. 33-49.
18. K. G. Tatterson, "Bioluminescent 'critters' make chip sensitive to air contaminants", <u>Biophotonics International</u>, (July/August, 1997) 33.
19. "No Payoff Without Delivery", <u>nature biotechnology</u>, 16 (February, 1998) lead editorial.

FUNCTIONAL COATINGS AS AN INTERESTING TOOL FOR INDUSTRIAL OPPORTUNITIES: DEVELOPMENT AND COMMERCIALIZATION

Helmut Schmidt
Institut für Neue Materialien GmbH
66123 Saarbrücken, Germany
Tel.: +49-681-9300-313/314
Fax: +49-681-9300-223
e-mail: schmidt@inm-gmbh.de

1. INTRODUCTION

After almost 50 years of sol-gel chemistry, the questions arise about the state of commercialization of sol-gel materials. As investigated by many authors [1-4] sol-gel materials show very interesting properties as well as processing advantages. The intriguing idea was to fabricate glasses and ceramics through a liquid route [5]. This idea was first patented by Geffcken and Berger [6] and commercialized by Schott glass works in the 70's. It could be shown that the densification of thin sol-gel films, even in high Tg systems as SiO_2/TiO_2 could be achieved at temperatures below the softening point of the coated glass panes. This technology was commercialized for architectural glazing and gave a strong input into the material science community, and many investigations have been carried out for the further development and exploitation of these basic findings.

This example of commercialization, however, already pointed out one way how sol-gel technologies can be successfully commercialized, namely by rising the value of a product for an already existing market. This value increase is substantial and actually also may open up new markets, as shown in the case of architectural glazing from Schott. For a successful introduction into the market, one has to combine chemistry, chemical processing, coating technology, application or production technology, marketing and sales. The involvement of several disciplines is an indispensible requirement for the commercialization, and a company has to have access to all these disciplines. This requires a reasonable company size including a profound chemical knowledge. This is one of the basic aspects of sol-gel commercialization, but the majority of material users, unfortunately, do not have these skills.

In this paper, the question how to utilize sol-gel chemistry in a broader way will be discussed and examples for successful utilization, especially involving academia, are presented. It has to be said that in most areas where sol-gel technologies are successfully used, large companies which already have chemical departments or have built up chemical departments are involved. It is also true that most of these companies use the sol-gel process as an in-house high performance tool for developing high-tech products. There are only a few examples in which chemical industry is producing materials to be used in down-stream processes. Examples are TiO_2 as a pigment, wash coats for automotive catalysts or sol-gel coated mica pigments with interference layers, especially as additives for automotive lacquers, or more recently, ITO slurries for the fabrication of conductive coatings for displays. In most cases, however, the in-house route is preferred, since the added value can be „harvested" through the final product.

2. BASIC MARKET CONSIDERATIONS

It is well known that the time to development and time to market of materials has become rather long. In the opinion of experts, these times are 10 years or more. This means that the development costs of materials are rather high and the risk is fully covered by the material developers. On the other hand, if one considers the added value from the material to the system, the material always covers the first or lowest step of the added value chain only.

This leads to a situation that for a successful marketing of sol-gel materials by selling these materials directly, relatively large market volumes have to be obtained, otherwise the development costs will not be paid back. This leads to the situation that the vast majority of interesting materials never has been commercialized due to market restrictions. If one now takes into consideration that for the commercialization of the „sol-gel raw materials" only the chemical industry can be considered, it becomes clear that only a few mass commodities have been successfully commercialized. Second, the number of companies who are equipped with chemical development facilities plus the down-stream processing to high-added value material is also very restricted. Typical examples are the electronic industry, sometimes the sensor industry, and, in a few cases, also the automotive industry and large suppliers. On the other hand, the structures in academia only allow the generation of knowledge in the specific field of the scientists in charge. The required, well-managed „vertical" interdisciplinarity is

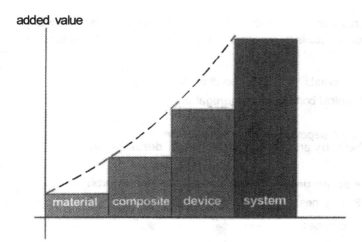

Fig. 1: Increase of added value from the material to the system

not organized in academia and one has to think about how to establish these structures. In the following, some technology platforms for sol-gel and related processes are given and it is shown, how they have been exploited for commercial use.

3. TECHNOLOGY BASIS
3.1. Fabrication of Nanoparticles

For the fabrication of nanoparticles, the sol-gel process has been choosen, since sols typically represents colloidal solutions with particle sizes in the nanometer range. The sol-gel process or the colloidal route to material is a route used by nature to form minerals like chalcedony or agate and known in industry for the fabrication of water glass since over 150 years. Many publications have been produced in the past [7-16]. Due to the large surface area and the strong particle to particle interaction, the sols have to be stabilized in order to prevent aggregation. Stern has shown that with electrostatically stabilized sol particles, stable and unstable regimes exist, depending on the particle to particle distance [17]. Other possibilities of stabilizing sols is the surface modification of sol particles during or after the sol formation, which can take place by the hydrolysis and condensation of alkoxides outside of the point of zero charge of the zeta potential curve, or by simple precipitation under similar conditions. Using surface modifiers, which bind to the colloidal particle surface by chemical bonds, results in several interesting options, which is schematically shown in figure 2. The stabilisation of the nanoparticles by electrostatic means only restricts the further use, since „leaving" Stern's stabilisation regime leads to uncontrolled aggregation (gel formation takes

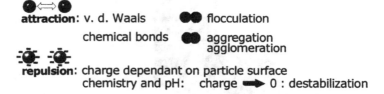

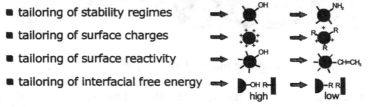

Fig. 2: Colloidal principles and advantages of surface modification of nanoparticulate sols

place). Gels, however, are rather difficult to process to compact materials, especially if no viscous flow sintering mechanisms can be employed for densification as it is the case for non-glassy systems. Silica is one of the exceptions and able to form large bodies as it has been shown for preforms by Lucent Technology.

The surface modification of colloidal particles by „small" molecules has several advantages. A large variety of surface modifiers can be used, which can be fixed on the surface by several types of chemical bonds, like hydrogen bridges (e.g. in aprotic solvents), complex, covalent or ionic bonds. If bifunctional molecules are used, the second function also may be of interest, since the surface chemistry of the sol particle can be influenced. Using this approach, the stability regime of the zeta potential curve, the amount of surface charges, the surface reactivity and the interfacial free energy between the particle surface and the matrix (liquids or polymer melts) can be tailored. This approach provides very interesting tools to fabricate „modified" sols, details of which are described elsewhere [18]. Based on these ideas, a fabrication process for modified sols and redispersible „dry sol-gel powders" has been established for medium size quantities [19].

3.2. Use of Specific Modification Techniques and Commercialized Products

3.2.1. Production of unreactive surfaces:
SiO_2 gels, in general, are brittle due to their three-dimensional crosslinking. In addition to this, they have a poor mechanical strength. To use SiO_2 sols as binders for glass fibers has been tried to be realized in vain for many times; the resulting glass fiber insulation materials are extremely brittle and can be crushed very easily by squeezing. In order to avoid these problems, silica sol has been surface modified by phenyl and methyl silanes in order to avoid particle to particle interaction for these modification techniques, which is described in detail elsewhere [20]. One of the effects of this surface modification is an improved relaxation behaviour of films produced from these modified sols. In a one step coating process films up to 15 μm thickness can be obtained without cracking after 500 ° densification on silica substrates [21]. Based on these results, it was expected that binders fabricated from these materials for glass fibers also show an improved flexibility and glass fiber insulating materials could have been developed based on this technique for industrial use. These glass fibers are flexible like polymer-bonded glass fibers and contain about 5 weight % of the modified binder. In opposition of the simple use of methyl or phenyl silanes, the modified SiO_2 leads to much more flexible systems due to the lower degree of three dimensional crosslinking [22]. The principle is shown in figure 3.

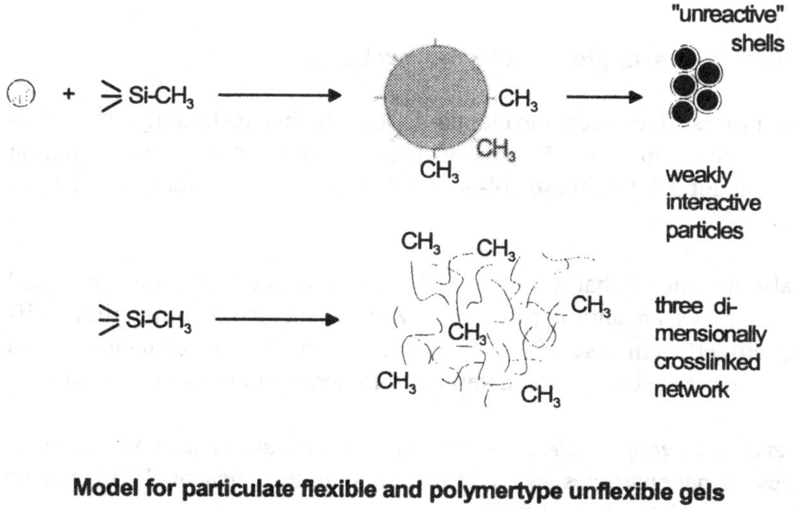

Model for particulate flexible and polymertype unflexible gels

Fig. 3: Principles of the surface modification of silica sols to be used as glass fiber binders (after [21]).

The flexibility of these fiber mats produced in an industrial process is shown in figure 4. The fibers are white, stable up to 500 °C over a long time and easily to be recycled. In case of fire, no toxic components are emitted.

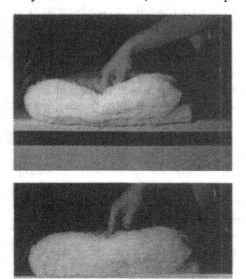

Fig. 4: New glass fiber mats produced by the new binder.

An industrial process has been developed based on this technology [23]. It is planned to produce up to about 70 tons of glass fibers per day at the beginning which requires about 3.5 t of these sol-gel binders per day and about 1000 t per year.

This binder also has shown that it can be used as an adhesive for natural fibers and wood, since it shows an antimicrobial behaviour, and, at the same time self-extinguishing properties in case of fire. The ability of the binding of natural fibers has been used to develop composite materials from hemp, straw and plywood [24].

3.2.2. Polymerizable nanoparticles: As shown elsewhere, the surface modification of many types of nanoparticles like alumina (boehmite, silica or TiO_2) can be carried out by appropriate silanes [25, 26].

In the case of boehmite, a process has been developed for coating the boehmite particles with an epoxy silane and to polymerize the epoxides by use of aluminum alkoxides or the boehmite particles as catalyst by themselves at temperatures of about 130 °C. Using this approach, very hard coatings have been fabricated with abrasion resistance values determined by a taber abrader test and measuring the haze values of about 2 % after 1000 cycles. These systems have been developed for the coating of polycarbonate for automotive glazing. The market expectations of this technology for a fraction of 10 % of the total glazing made from polycarbonate sums up to about 100 t of the sol-gel coating materials per year but the impact on the polycarbonate production is much more important. Further perspective in order to develop new designs for the cars through polycarbonate and to gain a larger market fraction leads to very interesting market volumes of 1000 and more tons per year, but the increased markets for polycarbonate are much more interesting, of course.

Another example for successful commercialization of the polymerizable nanoparticle technology is the fabrication of hard coatings for credit cards, driver's licenses and others. In this case, boehmite particles have been surface modified with methacryloxy silanes and the technology has been built up for off-set printing of these materials on plastic sheets for the fabrication of the described documents. The interesting part of this technology is that the laser damage threshold of these coatings at a thickness of several µm is high enough in order to permit laser writing through the coating into the plastic support. This permits the coating of the plastic sheets before the individual writing of the parts and facilitates the process substantially. About 150 million documents using these techniques are fabricated at present in Germany, but despite the fact that the material's value is rather small (several 100 kg per year), the added value of this technology is high. In figure 5 the principle of the process is shown.

3.3.3. Easy-to-clean technologies: The formation of gradient materials by sol-gel techniques has been proved with sol-gel hybrid materials using side chain perfluorinated silanes [27]. Due to thermodynamical effects, the fluorine is upconcentrated on the surface of the coating leading to a low surface free energy with an antiadhesion or, depending on application, with an easy-to-clean effect. The first successful commercialization of this type of coatings has been carried out in a common project together with nanogate, where the largest European group for sanitary ware and the world largest supplier for tile coating materials are involved. Sanitary ware is already very successfully available on the market and tile production is just introduced worldwide. The principles of the process are shown in figure 6. The market volume of this technology is large and is estimated

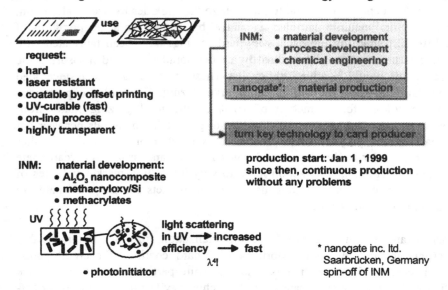

Fig. 5: Scheme of the development of a process for laser writable hard coated systems.

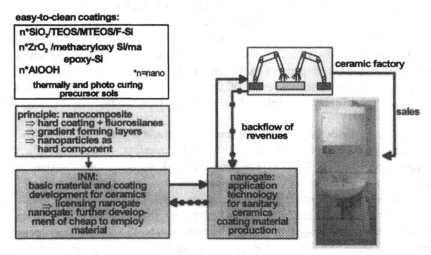

Fig. 6: Principles of the fabrication of easy-to-clean sanitary ware.

to be a couple of billion dollars per year worldwide, but the volume of the coating material again, is expected to be in the range of about 700 t per year.

3.3.4. *Catalytic systems for deodourisation*: The best way for the removal of odours is the decomposition or oxidation into simple components like water, carbon dioxide and nitrogen. The oxidation catalysis can be carried out with finely derived nobel metal on carriers, but also with specific transition metal oxides. However, the activity of transition metals is rather low compared to nobel metals. On the other side, nobel metal catalysts are expensive and in most cases rather difficult to be produced in desired shapes for specific applications. For these reasons, sol-gel binders have been used for binding mixtures of µm-sized powders of cobalt oxides, mangane oxides and cerium oxides. The idea was to produce open structures with large transport channels for a fast gas exchange and to reduce the operating temperatures down to about 300 °C. In figure 7 the microstructure of the catalyst bonded with surface modified sol-gel SiO_2 and sol-gel aluminum oxide as binders is shown. The technique permits to produce layers up to several hundred µm and in thickness on any type of substrates, e. g. ceramic honeycombs

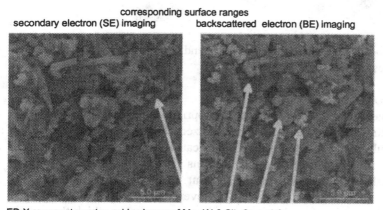

Fig. 7: SEM micrographs of the described oxide catalysts

or even on stainless steel. The resulting catalyst has an activity of about 10 times of the state of the art and has been commercialized together with one of the world's largest producers of kitchen stoves to remove the smell of cooked and roasted food. This technique provides an improvement of the competitiveness of the company and the catalytic material is produced by nanogate company in the quantity of several tons per year at present. In figure 8 the whole process is shown schematically. Again, the sol-gel materials produced even in small quantities improve competitiveness substantially.

- **objective:** unspecific oxidation catalysts for odor removal in kitchen
- **strategy :** state of the art compositions + sol-gel binders
  - micro structure
  - thick coatings ($\leq$300 μm)
  - viscosity
  - curability
- **material development**

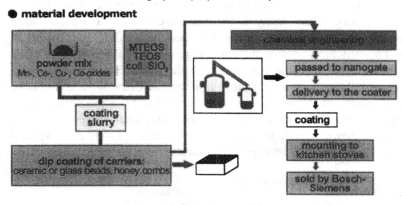

Fig. 8: Scheme of the process of the fabrication and commercialization of deodourisation catalysts.

7. CONCLUSIONS

The few examples show that by using the appropriate approach, sol-gel materials can be produced and successfully commercialized. However, it is necessary to obay certain rules. One of the rules is indicated as the so-called „vertical integration" which is shown in figure 9. It means that it is not sufficient only to develop materials with excellent properties, but also to provide a technology development on the academia level. This, however, means that the R&D partner has to collaborate in this vertical integration down to a level at which the partner is able to take up the development. This depends strongly on the abilities and the skills of the partners and especially on their size. The requirement for vertical interdisciplinarity is shown schematically in figure 10. As one can conclude from this figure, there are only a few technologies which can be transferred to industry in the basic science state. If one wants to transfer technologies either to smaller companies or to companies not familar with the sol-gel techniques, one has to process the technology down to production, and if production technology also is developed, small and medium sized enterprises become extremely interested in

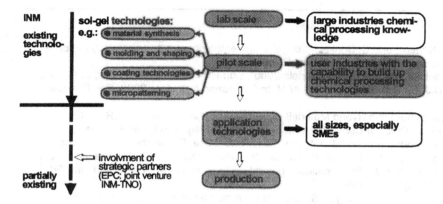

Fig. 9: The vertical route for the successful strategy for the commercialization of sol-gel products.

the new technologies. Summarizing it is to say that sol-gel techniques are a powerful tool for producing and developing new materials but due to the lack of vertical integration in R&D centers, only a small percentage of the developments are transferred into the market.

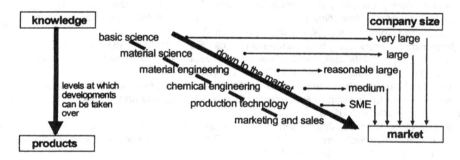

academica: in most cases only one discipline at the same time
required: interdisciplinary R+D organization with vertical interdiciplinarity

Fig. 10: Strategy and steps „down to the market"

REFERENCES:

[1] G.W. Scherer and C. Brinker, "Sol-gel science: the physics and chemistry of sol-gel processing" *Academic Press San Diego*, 1990.
[2] "Organic/Inorganic Hybrid Materials‹2000", Ed.: R.M. Laine, C. Sanchez, E. Giannelis, and C.J. Brinker, *Proc. Of Conf. of Material Research Society*, 2000, San Francisco, USA
[3] "Organic/Inorganic Hybrid Materials II", Ed.: L.C. Klein, L. Francis, M.R. DeGuire and J.E. Mark, *Proc. Of Conf. of Material Research Society*, 1999, San Francisco, USA
[4] "Organic/Inorganic Hybrid Materials", Ed.: R.M. Laine, C. Sanchez, E. Giannelis and C.J. Brinker, *Proc. Of Conf. of Material Research Society*, 1998, San Francisco, USA
[5] H. Dislich, "New routes to multicomponent oxide glasses", *Angew. Chem., Int. Ed. Engl.* **10** (6) 363-70 (1971).
[6] W. Geffcken and E. Berger „Schutzschichten aufglasen mit Schutzmantel" DD 963/75568 (1943)
[7] R. Roy, "Materials preparation and characterization research", *U.S. Govt. Res. Develop. Rept.*, **41**(20) 182 (1966)
[8] "Eurogel 92: Practical applications and innovative materials by sol-gel processing," Ed.: S. Vilminot, *Proc. of 3rd European Conference on Sol-Gel Technology*, 1992 Colmar, France
[9] "Eurogel 91 : Progress in research and development of processes and products from sols and gels", Ed.: S. Vilminot, H. Schmidt and R.Nass, *Proc of 2nd European Conference on Sol-Gel Technology*, 1991, Saarbrücken, Germany
[10] "Ultrastructure processing of advanced materials", Ed.: D. R. Uhlmann and D. R. Ulrich, *Proc. of 4th International Conference on Ultrastructure Processing of Ceramics, Glasses and Composites*, 1989, Tucson, AZ (USA)
[11] "Ultrastructure Processing of advanced ceramics"; Ed.: J. D. Mackenzie and D. R. Ulrich, *Proc. of 3rd International Conference on Ultrastructure Processing of Ceramics, Glasses and Composites*, 1987, San Diego, CA (USA)
[12] "Science of ceramic chemical processing"; Ed.: L. L. Hench, *Proc. of 2nd International Conference on Ultrastructure Processing of Ceramics, Glasses and Composites*, 1985, Gainesville, FL (USA)
[13] "Ultrastructure of Processing of Ceramics, Glasses and Composites ", Ed.: L. L. Hench and D. R. Ulrich, *Proc. of Ultrastructure Processing of Ceramics, Glasses and Composites*, 1983, Gainesville, Fl, USA
[14] "Better Ceramics Through Chemistry VII: Organic/Inorganic Hybrid Materials", Ed.: B.K. Coltrain, C. Sanchez, D.W. Schaefer and G. L. Wilkes, *Proc. of Conf. of Material Research Society*, 1996, San Francisco, USA
[15] "Better Ceramics Through Chemistry VI", Ed.: A.K. Cheetham, C.J. Brinker, M. L. Mecartney and C. Sanchez, *Proc. of Conf. of Material Research Society*, 1994, San Francisco, USA
[16] "Better Ceramics Through Chemistry V", Ed.: M. J. Hampden-Smith, W. G. Klemperer and C. J. Brinker, *Proc. of Conf. of Material Research Society*, 1992, San Francisco, USA
[17] O. Stern, *Z. Elektrochem.* **30** 508 (1924)

[18] H. Schmidt, "Relevance of sol-gel methods for synthesis of fine particles," *Kona powder and particle* **14** 92-103 (1996).

[19] H. Schmidt and R. Nonninger, „Chemical Routes to Nanoparticles: Synthesis, Processing and Application", *Proc. of Fine, Ultrafine and Nano Powders '98* (1998)

[20] H. Schmidt, "Sol-Gel Derived Nanoparticles as Inorganic Phases in Polymer-type Matrices" *Proc of Journal of Sol-Gel Science and Technology), Yokohama* (1999) in print

[21] M. Mennig, G. Jonschker, and H. Schmidt, "Sol-gel derived thick coatings and their thermomechanical and optical properties", *Proc. of SPIE,* **1758** 125-134 (1992).

[22] G. Jonschker, *Ph. D. Thesis*, University of Saarland, Saarbrucken, Germany (1998)

[23] Pfleiderer: insulating materials company

[24] G. Jonschker, M. Mennig and H. Schmidt, "Verbundwerkstoffe," DE 19647368 (1996)

[25] E. Arpac, H. Krug, P. Müller, P.W. Oliveira, H. Schmidt, S. Sepeur and B. Werner, "Nanostrukturierte Formkörper und Schichten sowie Verfahren zu deren Herstellung," DE 19746885 (1997)

[26] E. Geiter and H. Schmidt, "Verwendung von nanoskaligen Metalloxid-Teilchen als Polymerisationskatalysatoren," DE 19726829 (1997)

[27] H. Schmidt, R. Kasemann and S. Brueck, "Beschichtungszusammensetzungen auf der Basis von fluorhaltigen anorganischen Polykondensaten, deren Herstellung und deren Verwendung," DE 4118184 (1991)

SOL-GEL COMMERCIALIZATION IN JAPAN

Sumio Sakka
Fukui University of Technology
3-6-1 Gakuen
Fukui, 910-8505, Japan

ABSTRACT
Sol-gel materials of various shapes, microstructures and properties have been manufactured in Japanese industries. The conditions for the successful commercialization are discussed, based on the list of the industrial products. It is shown that materials fabricated only by the sol-gel method, innovative finding in sol-gel technology, application of sol-gel process without disturbing original manufacturing processes, subjects related to environmental problems and subjects involved in the current trend lead to the successful commercialization of the sol-gel product.

INTRODUCTION
Almost thirty years have passed since Dislich [1] prepared a transparent borosilicate glass lense by the sol-gel method, showing the potentiality of the method to fabricate practically useful materials. A great number of excellent research papers on the basic science of the sol-gel processes have been published since then. On the other hand, there is a feeling among sol-gel people that the number of sol-gel products is not so big as assumed from that of basic researches. This situation may be partly caused by people's unawareness of the sol-gel products and even the sol-gel method.
In order to improve this unhappy state, I introduced sol-gel industrial products and potential application of the sol-gel method to advanced materials in 1998 at the Sol-Gel Symposium at the Annual Meeting of American Ceramic Society [2].
In this paper representative sol-gel products in Japan will be listed up again. A small number of interesting sol-gel products which are commercialized afterward are added. Brief explanation will be given. However, a stress is laid on the discussion about the reason for the successful commercialization.

ADVANTAGES OF SOL-GEL METHOD
In sol-gel method, materials are prepared through the gelation of solutions or sols. Accordingly, the sol-gel method is characterized by the low processing temperature. When the resulting gel is regarded as the final product, the processing temperature may be near room temperature, and lower than 200°C, at the highest. This enables to make organic-inorganic hybrid materials and nanocomposites,, because organic materials or residues remain without decomposition. Even when

the gel has to be heated for convertion to non-porous glass or ceramic, the sintering temperature may be much lower than that for sintering conventional powder compacts. Preparation of materials through the gelation of solutions also makes it possible to prepare shaped materials, such as bulk bodies, films and fibers without the use of powder processing. Also, materials of various microstructures can be produced.

Industrial products fabricated based on those characteristics of sol-gel method are shown and explained in the followings.

SOL-GEL INDUSTRIAL PRODUCTS IN JAPAN

Bulk Materials: Table 1 lists up important examples of sol-gel industrial products of bulk materials developed in Japan.

Table I. Bulk materials developed in Japan

	Year	Material	Author	Manufacturer
(1)	1986	Machinable ceramics	Hamasaki [3]	Mitsui Mining Materials
(2)	1989	Alumina tape	Shinohara [4]	Mitsubishi Material
(3)	1992	Graded index lens	Yamane [5]	Olympus Camera
(4)	1996	Porous silica for chromatography	Nakanishi [6]	Merck (Germany)
(5)	1997	Aerogel	Yokoyama [7]	Matsushita Electric Works
(6)	1998	Wood-inorganic composite	Saka [8]	Shin-Etsu Chemicals

(1) Machinable ceramics

The machinable mica ceramics can be cut by conventional metallic saw. Sol-gel prepared powder for machinable mica ceramics could be sintered at temperatures 100°C lower than the conventional powder compact [3]. They are used as insulating material for precision machining and substrate for electronic parts.

(2) Alumina tape

Alumina tapes were prepared by firing green sheets formed from very viscous sols by doctor blade technique at 1330°C [4]. Dense alumina tapes, 30-100 μm thick, are used as ceramics for medical instrument, microphone vibrating film and microelectronics.

(3) Graded index lenses.

Graded index lenses have little aberration and used as lenses for copying-machine, connection of optical fibers, camera and so on. These lenses are-prepared by incorporation of high refractive index oxide into porous gels [5]. Sol-gel method is very suitable for the production of graded index rod lenses, because it gives porous gels.

(4) Porous silica rod for chromatography.

Usually, gels are microporous and are used as filters and catalyst supports. Nakanishi [6] developed a new sol-gel method for making porous silica gels of double pore system from silicon alkoxide-water soluble organic polymer compositions. Macropores and mesopores are formed as a result of concurrent gelation and phase separation due to spinodal decomposition. In these materials, the size of macropores and that of mesopores can be well controlled independently. Macropores are used for transport of substrate and mesopores are used for interaction of substrate molecules with pore surfaces. The porous monolith rods of about 10 mm diameter amd 100 mm length are used for liquid chromatography.

(5) Aerogel
In 1997, Matsushita Electric Company announced commercialization of bulk aerogels [7]. They developed hydrophobic, transparent aerogel plates of 10 mm in thickness and 10 cm x 10 cm in area. The highest quality aerogel, that is, the most transparent aerogel is for the Cerenkov effect measurement in observation of high energy particles in physics. Aerogels of lower prices are for insulation.

(6) Wood-inorganic composites
Most of Japanese individual houses are made from wood, which needs protection against fire, rot and insect. For protection of wood, Saka and colleagues [8] have been applying sol-gel process to prepare wood-inorganic composites. Originally, moisture-conditioned wood was soaked in the silicon alkoxide-alcohol-acetic acid solution at an ambient temperature under reduced pressure. The resulting SiO_2 wood-inorganic composites were found to have higher fire-resisting properties. Wood-inorganic composites using SiO_2-P_2O_5, SiO_2-B_2O_5 and SiO_2-P_2O_5-B_2O_5 compositions showed higher fire-resistance. Addition of an antimicrobial alkoxysilane (3-(trimethoxysilyl) propyl dimethyl actadecyl ammonium chloride) and a water repellent compound (2-heptadecafluorooctylethylrimethoxysilane) to the sol-gel solution gave the long-lasting antimicrobial property to the wood by suppressing white-rot and brown-rot fungi. A solution for giving anti-insect property to wood was also produced.

The solutions for preparing wood-inorganic composites are commercialized by Shinetsu Chemical Industry Company [9].

Coating Films: In Japan, a large number of sol-gel coating films are commercialized, as shown in Table II.

1) HUD combiner for automobiles.
An HUD (Head Up Display) combiner is a light reflecting SiO_2-TiO_2 coating film, about 200 nm thick, deposited on the windshield of automobiles [10]. With this system the driver can see the speed of the car without looking down on the panel board. For instance, the speed is displayed on the windshield, . This film shows more than 20 % reflectivity, which permits seeing the displayed signs clearly, and more than 70 % transmittance, which agrees to the legal regulation concerning the visibility of the windshield. It should be noted that this triggered the commercialization of the sol-gel coating films as applied to automobile windows.

(2) Protecting film on steel.

Table II Coating films developed in Japan

	Year	Material	Author	Manufacturer
(1)	1988	HUD combiner for automobiles	Makita [10]	Central Glass
(2)	1989	Protecting film on steel	Izumi [11]	Nisshin Steel
(3)	1990	Selectively absorbing coating for CRT	Itoh [12]	Toshiba
(4)	1994	Antireflection coating for CRT	Hayama [13]	Matsushita Electronics
(5)	1992	Water-repellent coating for automobiles	Yamasaki [14] Morimoto [15]	Central Glass Asahi Glass
(6)	1996	UV-shielding coating for automobiles:	Morimoto [16]	Asahi Glass
(7)	1996	Protecting coating for building materials	Yamada [17]	JSR
(8)	1998	Transparent photocatalyst	Fujishima [18] Watanabe [19]	Toto [20]
(9)	1999	Film for recyclable colored bottles	Nakazumi [21] Minami[22]	

Izumi et al. [18] developed protecting (oxidation-resistant and corrosion-resistant) coating films based on SiO_2. A part of $Si(OC_2H_5)_4$ as source of SiO_2 was replaced by $CH_3Si(OC_2H_5)_3$, in order to give the coating a flexibility necessary for bending the steel sheets after coating without crack formation.

(3) Selectively absorbing coating for CRT.
In order to improve clear vision of color television picture, Itoh et al. of Toshiba [12] coated the television panel glass with a selectively absorbing coating film. SiO_2-ZrO_2 coating film containing organic pigment which absorbs the disturbing lights, achieving remarkable increase in contrast, was successfully applied to the panel glass. The success of this film induced people to apply sol-gel coating to color television glass.

(4) Antireflection coating film.
Hayama et al. of Matsushita Electronic Company [13] coated the panel glass with a low-reflection, non-glare and antistatic three layer film. The lower layer is a conducting high refractive index layer of SiO_2-TiO_2 system with dispersed Sb-doped SnO_2 particles. The intermediate layer is low refractive index SiO_2 layer. The upper layer is a low refractive index layer with uneven structure. Accordingly, the coating film gives a low-reflection and non-glare effect and antistatic effect.

(5) Water-repellent coating for automobiles.
Water-repellent coatings consisting of SiO_2 doped with fluoroalkyltrimethoxysilane ($CF_3(CF_2)_7CH_2CH_2Si(OCH_3)_3$), a water-repellent fluorine compound, were applied to automobile windows and mirrors by Central Glass[14] and Asahi Glass[15]. These windows and mirrors assure clear sight on rainy days.

(6) UV shielding coating for automobiles.
A UV shielding glass for automobiles was developed by sol-gel method in Asahi Glass Company [16], in order to protect passengers from exposure to UV lights.

The coating film consists of two layers: an upper layer of CeO_2-TiO_2 complex oxide absorbing ultraviolet rays and a lower layer of low refractive index. The lower layer is used to avoid unfavorable yellow coloring of the coated window due to the selective light reflection by the upper layer.

(7) Protecting coating for building materials.

JSR Corporation commercialized sols for organic-inorganic hybrid thick coating films for building walls and tiles[17]. The final product of thick, transparent coating film consists of polyorganosiloxanes containing colloidal silica or alumina particles. In order to prepare the sol, methyl-triethoxysilanes are hydrolyzed and polycondensed under the effect of polymerization catalyst of aluminum ethylacetoacetate ($Al(CH_3COCH_2COOC_2H_5)_3$). When the molecular weight of polyorganosiloxanes reaches a given value, acetylacetone is added to the solution, in order to suppress the further polymerization which leads to unfavorable gelation before use. Acetylacetone causes ligand exchange, changing aluminum ethyl-acetoacetate into aluminum acetyl-acetonato ($Al(CH_3COCH_2COCH_3)_3$), which suppresses further polymerization. This makes the pot life of the sol very long (longer than 4 months at 45 °C). The resultant sol is applied to the building walls as hard, durable coatings, thicker than 20 μm. The success of sol-gel coating for building walls indicates that application of sol-gel processing should not be limited to high technology materials.

(8) Transparent photocatalyst.

When the anatase formed TiO_2 is irradiated with UV light in presence of water, it decomposes many organic materials into CO_2 and H_2O [18,19]. Toto Company developed the photocatalyst sol-gel coatings consisting of SiO_2-based films containing TiO_2 fine particles[20]. These coatings are used for deodorization in rooms, self-cleaning of building walls and glass lamps in tunnels and removal of NOx on highways. Many other companies are working on photocatalyst coatings.

(9) Film for recyclable colored bottles.

Nakazumi et al.[21] developed highly durable SiO_2-based colored coating films for colored glass bottles which can be effectively recycled. Organic pigments are used. So, the bottles become colorless upon heating at 400°C. It should be remarked that this work was carried out by cooperation between university and bottle makers [22].

Fibers, Particles and Powders: Sol-gel method is applied to fabrication of high performance inorganic fibers, mono-sized particles and powders for sintering ceramics. These are listed in Table III.

(1) Alumina fiber.

Heat-resistant alumina fibers doped with 15 % silica were commercialized at the early days of the present sol-gel method. Sumitomo Chemical Company produced alumina fiber by continuously drawing from the solution containing polymerized aluminoxanes [23].

Table III Fibers, particles and powders

	Year	Material	Author	Manufacturer
(1)	1974	Alumina fibers	Horikiri [23]	Sumitomo Chemical
(2)	1976	SiC fibers	Yajima [24]	Nippon Carbon
(3)	1988	Silica fibers	Taneda [25]	Asahi Glass
(4)	1996	Silica spheres	Adachi [26]	Ube-Nitto Chemical
(5)	1987	Mullite powder	Ismail [27]	Taiheiyo Cement
(6)	1982	Alumina powder	Takeuchi [28]	Sumitomo Chemical
(7)	1993	UV cut silica flakes	Yokoi [29]	Nipponn Sheet Glass

(2) SiC fibers.

SiC fiber named Nicalon® is a very famous and important sol-gel product [24]. The SiC fibers are produced by drawing from polycarbosilane melt at 400°C and heating the resultant gel fibers to 1300°C. The fibers, consisting of β-SiC fine particles, show a high tensile strength and Young' modulus.

(3) Silica fibers.

Silica fibers were produced by drawing from sols containing $Si(OC_2H_5)_4$ in Asahi Glass Company [25]. The viscous sol becomes drawable when the sol contains long-shaped polymerized fibrous particles. The heat-resistant inorganic fibers are used as supports of high temperature oxidation catalysts. Unfortunately, the production of silica fibers was stopped later.

(4) Silica spheric particles.

Liquid crystal display is very popular. In order to keep the liquid crystal layer at a given thickness (several microns to ten microns), spacers are needed. Originally, high polymer spheres were employed for spacers. Afterwards, hard spacers were needed, and silica microspheres which have a diameter larger than several microns were developed in Ube-Nitto Chemical Company [26] and other companies.

(5) Mullite powder.

Gel powder[27] prepared from the $3Al_2O_3 \cdot 2SiO_2$ sol consisting of boehmite sol and silica suspension is inverted to mullite crystals at 1400°C and forms a very dense ceramic having a density higher than 98 % of theoretical density at 1650°C. The mullite ceramics thus prepared are characteristic of high mechanical strength. This powder is commercialized.

(6) Alumina powder.

Alumina powder of Sumitomo Chemical Company [28] prepared by sol-gel method consists of round particles with a very sharp size distribution. This powder is an excellent starting material for fabricating sintered ceramics.

(7) UV cut silica flakes (TSG Flakes).

TSG Flakes developed in Nippon Sheet Glass are amorphous silica flakes with submicron thickness in which titania fine particles of 30-60 nm in diameter are dispersed [29]. They effectively cut UV light harmful to skin, while they are transparent. They are incorporated into UV shielding cosmetics

DISCUSSION ON SUCCESSFUL COMMERCIALIZATION

Naturally, manufacturing of the products listed in Table I - III utilizes the advantages of sol-gel method. Actually, many other technological progresses and condtions affect the success of the manufacturing of a sol-gel product. This is explained on the basis of Table IV below. Some of such reasons for the success and related examples of the commercialized product are shown in Table IV and are discussed.

Table IV Reasons for successful commercialization

Reason	Example
(1) Fabrication of new materials which can be prepared only by sol-gel method.	Aerogel (Table I (5)). Porous silica for chromatoraphy (Table I (4)). Silica spheres (Table III (4))
(2) Technological innovation in sol-gel.	Protecting coating for building materials (Table II (7)).
(3) Incorporation of sol-gel process without disturbing conventional fabrication processes.	Selectively absorbing coating for CRT (Table II (3)).
(4) The subjects related to environmental problems	Transparent photocatalyst (Table II (8)). Recyclable colored bottles (Table II (9)).
(5) Current trend	Coatings for CRT (Table II (3) (4)) Coatings for automobiles (Table II (5) (6))

(1) Materials fabricated only by sol-gel method.

Porous silica rods for chromatography (Table I (4)) made by sol-gel method are excellent elements for high performance liquid chromatography [6]. The size, size distribution and pore content can separately be controlled both for fine pores for adsorption and desorption and large pores for transport of liquid substrate. In this respect the sol-gel rod is superior to the conventional chromatographic element, which is a silica powder compact.

Obviously, aerogel (Table I (5)) can be prepared only by sol-gel method [7].

Silica spheric particles (Table III (4)), used as spacers for liquid crystal display devices, are microparticles of several to 10 μm in diameter [26]. A very sharp size distribution is needed. Only sol-gel method can produce such particles.

(2) Technological innovation

Commercialization of protecting coating for building materials, such as walls and tiles (Table II (7)) is based on a technologicals innovation in sol-gel method. JSR sells canned sols to companies manufacturing building material. The engineers in JSR found a new aluminate catalyst for hydrolysis and polymerization of the starting solution, which is active until the reaction reaches a desired point and then can easily be made inactive for a long pot life. This is the main reaction for the success of commercialization of this product.

(3) Process without disturbing conventional processes.

When selectively absorbing coating for CRT (Table II (3)) is fabricated, coating is carried out after all other processings are finished. Therefore, the coating process does not disturb the conventional manufacturing processes. This makes it easy to add the coating to the conventional manufacturing process.

(4) Subject related to environmental problems.

Transparent photocatalyst (Table II (8)) and recyclable colored bottles (Table II (9)) are related to environmental problems. This is one of the reasons for the successful commercialization of those sol-gel products.

(5) Current trend.

Electronics companies are currently trying to improve the visibility and color contrast of television glass. Coatings for CRT (Table II (3) and (4)) have been commercialized based on the current trend. It is considered that coatings for automobiles (Table II (5) and (6)) have been commercialized based on the current trend of donating automobile windows with UV-shielding and water-repellent properties.

SUMMARY

It was shown that many materials prepared by sol-gel method are commercialized. The reasons for the sucessful commercialization were discussed. I hope that more industrial products will appear in the future.

REFERENCES

1. H.Dislich, "New Routes to Multicomponent Oxide Glasses", Angewandte Chemie, International Edition, 10, 363-370 (1971).
2. Sumio Sakka, "History of Sol-Gel Technology in Japan", Sol-Gel Synthesis and Processing, Ceramic Transactions, Vol. 95, ed.by S.Komarneni, S.Sakka, P.P.Phulé and R.M.Laine (1998) p.37-49.
3. T.Hamasaki, K.Eguchi, Y.Koyanagi, A.Matsumoto, T.Utsunomiya and K.Koba, "Preparation and Characterization of Machinable Mica Glass-Ceramics by the Sol-Gel Process," Journal of the American Ceramic Society, 71, 1120-1124 (1988) ; Toshio Hamasaki, "Development of Machinable Mica Glass-Ceramics by the Sol-Gel Process", NEW GLASS, 4 [4] 38-49 (1990).
4. Y.Shinohara, M.Hirai and M.Ono, "Applications of Ultrathin Alumina Substrates," pp. 481-487 in Advances in Ceramics, Vol 26, The American Ceramic Society, 1989.
5. M.Yamane and M.Inami, "Variable Refractive Index Systems by Sol-Gel Process," Journal of Non-Crystalline Solids, 147 & 148, 606-613 (1992).
6. K.Nakanishi, "Synthesis and Application of Double Pore Silica via Sol-Gel Route," New Ceramics, 9 [8] 68-72 (1996). (in Japanese).

7. H.Yokogawa and M.Yokoyama, "Hydrophobic Silica Aerogels," Journal of Non-Crystalline Solids, 186, 23-29 (1995).

8. F.Tanno, S.Saka and K.Takabe, "Antimicrobial TMSAC-led Wood-Inorganic Composites Prepared by the Sol-Gel Process", Materials Science Research International, 3 [3] 137-142 (1997).

9. Shinetsu Chemical Industry Company Catalogue, Oligomer for Modification of Wood X-40-2273; Antibacterial, Antifungi Reagent X-24-9179 (Dec.7, 1998). (in Japanese).

10. A.Hattori, K.Makita and S.Okabayashi, " Development of HUD Combiner for Automotive Windshield Application," pp. 272-282 in SPIE Vol 1168, Current Developments in Optical Engineering of Commercial Optics, 1989.

11. M.Murakami, K.Izumi, T.Deguchi, A.Morita, N.Tohge and T.Minami, "SiO_2 Coating on Stainless Steel Sheets from $CH_3Si(OC_2H_5)_3$", Journal of Ceramic Society of Japan, 97, 91-94 (1989). (in Japanese).

12. T.Itoh, H.Matsuda and K.Shimizu, "Black Enhance Color Picture Tube," Toshiba Review, 45 [10] 831-834 (1990). (in Japanese).

13. H.Hayama, T.Aoyama, T.Utsumi, Y.Miura, A.Suzuki and K.Ishiai, "Anti-Reflection and Anti-Static Coating for CRT," , National Technical Report ,40 [1] 90-96 (1994).

14. S.Yamasaki, H.Inaba, H.Sakai, M.Tatsumisago, N.Tohge and T.Minami, "Water-Repellent Coatings on Glass Substrates by the Sol-Gel Process", pp. 291-295 in Boletin de la Sociedad Espanola de Ceramica y Vidrio, 31-C, Vol 7 (Proceedings of XVI International Congress on Glass), 1992.

15. F.Gunji, T.Yoneda and T.Morimoto, "Novel Functioning of Glass by Wet Coating Process: Part (I) Water-Repellent Glass," New Glass 11 [4] 49-56 (1996). (in Japanese).

16. H.Morimoto, H.Tomonaga and A.Mitani, "Ultraviolet Ray Absorbing Coatings on Glass for Automobiles," Coatings on Glass 1998, Ed. H.Pulker, H.Schmidt, M.A.Aegerter, Elsevier Science B.V. (1999) p.95-99.

17. Y.Yoshida, H.Hanaoka, M.Nagata, T.Sakagami and K.Yamada, "Development of Organic-Inorganic Hybrid Coating Agent by Sol-Gel Method", Submitted to Journal of Japan Chemical Society (1988). (in Japanese); T. Sakagami, "Organic-Inorganic Hybrid Coating", Industrial Materials, 46 [8] 57-61 (1998). (in Japanese).

18. A.Fujishima, "Photocatalyst Superhydrophilic Technology and its Application", Optical Technology Contact, 35[11]626-632 (1997). (in Japanese).

19. T.Watanabe, "Super-Hydrophilic TiO_2 Photocatalyst and its Applications", Ceramics, Japan, 31[10]837-841 (1996).

20. Toto Catalogue, "Toto Hydrotect on Photocatalyst Superhydrophilic Technology", Toto, Kimura 2-8-1, Chigasaki City, Kanagawa-Ken, 253-8577, Japan. (in Japanese).

21. H.Nakazumi, K.Ishii, Y.Sakashita, T.Ikai, K.Tadanaga, T.Minami, M.Ueda, M.Funato, H.Kanazawa, K.Nakatsukawa and Y.Sakurai, "Color Coating for Recyklable Glass Bottle by the Sol-Gel Method Using Organic Colorants", Coatings on Glass 1998, Ed. H.Pulker, H.Schmidt and M.A.Aegerter, Elsevier Science B.V. (1999) p.114-119.

22. Brochure delivered to the participants of the 10th International Workshop on Glasses, Ceramics, Hybrids and Nanocomposites from Gels, "Sol-Gel Coated Wine Bottle", T. Minami, Osaka Prefecture Univeraity (representative of the Project) (1999).

23. S.Horikiri, "Alumina Fibers and their Application", Ceramics (Japan), 19, 194-200 (1984). (in Japanese).

24. S.Yajima, K.Okamoto, J.Hayashi and M.Omori, "Synthesis of Continuous SiC Fiber with High Tensile Strength", J.Am.Ceram.Soc., 59, 324-327 (1976).

25. N.Taneda, K.Matsusaki, T.Arai, T.Mukaiyama and M.Ikemura, "Properties and Applications of Silica Fiber by Sol-Gel Process", Reports of Research Laboratory of Asahi Glass Company, 38, 309-318 (1988).

26. K.Toda and T.Adachi, "Silica Spacers".Electronics Materials, June 1996 extra issue, 54-58 (1996). (in Japanese).

27. M.G.M.U.Ismail, Z.Nakai and S.Somiya, "Microstructure and Mechanical Properties of Mullite Prepared by the Sol-Gel Method," Journal of the American Ceramic Society, 70, C7-C8 (1987).

28. Y. Takeuchi, H.Umezaki and H.Kadokura, "High Purity Alumina Derived from Aluminum Alkoxide", Sumitomo-kagaku, 1993-I, p.4-14. (in Japanese). ;
K.Yamada, "Development of Ceramic Materials by Sol-Gel Process", NEW GLASS, 4 [4] 50-55 (1990). (in Japanese).

29. Nippon Sheet Glass Brochure, "UV Cut Silica Flakes(TSG Flakes), Nippon Sheet Glass Co.Ltd., NGF Company, Takachaya, Tsu, Mie-Ken, 514-0817, Japan. (in Japanese).

MOLECULAR TEMPLATED SOL-GEL SYNTHESIS OF NANOPOROUS DIELECTRIC FILMS

Suresh Baskaran, Jun Liu,* Xiaohong Li, Glen E. Fryxell, Nathan Kohler, Christopher A. Coyle, Jerome C. Birnbaum, and Glen Dunham
Pacific Northwest National Laboratory, Battelle Boulevard, Richland, WA 99352

ABSTRACT

The molecular-templated approach using surfactant micelles as pore-forming agents provides an effective route to prepare high-porosity (up to 65 volume %) nanoporous dielectric films with a uniform pore diameter smaller than 5 nm. To accomplish this, we prepared a homogeneous solution by mixing silica alkoxide and surfactant in an acidic aqueous environment. During spin coating, the solvent was rapidly removed, and the surfactant molecules formed micellar aggregates. After the surfactant was removed, a highly porous silica film was produced with controlled porosity and pore sizes. When proper dehydroxylation treatments were applied, stable, low-dielectric constants between 1.8 and 2.2 were obtained. The ability to increase and control the porosity and to optimize the materials parameters makes this approach attractive for next-generation ultralow dielectrics.

INTRODUCTION

The semiconductor industry is currently targeting new intermetal dielectric films with k <2.5 for advanced interconnects, and it is anticipated that as the packing density of metal lines on the semiconductors continues to increase, intermetal dielectric films with k <2.0 will be soon required. Porous silica films with nanometer-scale porosity are potentially useful as intermetal materials with a low dielectric constant in advanced semiconductor devices. Nanoporous silica films have been synthesized on semiconductor substrates from solution precursors in two general ways: (1) by an "aerogel or xerogel" process [1-4] in which random porosity (where the pore size is typically much larger than 5 nm) is introduced by controlled gelation and removal of an alcohol-type solvent or co-solvent and (2) by a surfactant-templating process [5-7] in which nanoporosity is formed in a spin-coated [8-11] or dip-coated film [12,13] by removal of a surfactant.

In addition to their nominal dielectric constant and mechanical integrity, intermetal dielectric films must meet some key physical/engineering criteria. The

---

To the extent authorized under the laws of the United States of America, all copyright interests in this publication are the property of The American Ceramic Society. Any duplication, reproduction, or republication of this publication or any part thereof, without the express written consent of The American Ceramic Society or fee paid to the Copyright Clearance Center, is prohibited.

films will be used in building six to nine levels of dielectric with inlaid metal over the silicon device. The principal requirements in terms of physical properties are good control of film thickness, high thickness uniformity across large wafers, and low surface roughness. Thickness must be controlled accurately in the 0.5–1.0 micrometer range. Spin coating is the most convenient and economical manufacturing process for depositing dielectric films from solutions on large-area wafers. Surface roughness, especially in terms of radial striations (which slow the throughput of wafers in manufacturing), can be a serious problem in spin coating, requiring careful control of atmospheric conditions over the wafer. Striations are thick ridges radiating from the center of the substrate in spin-coated films. As both the sizes of the interconnect features and the thickness of the layers decrease, the uniformity of the film thickness and the planarity become even more critical.

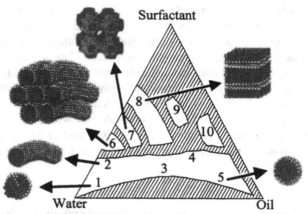

Figure 1. A simplified surfactant-oil-water phase diagram showing the ordered phases.[14,15] (1) Spherical micelles. (2) Cylindrical micelles. (3) Microemulsion. (4) Inverse cylindrical micelles. (5) Inverse spherical micelles. (6) Hexagonal phase. (7) Cubic phase. (8) Lamella phase. (9) Inverse cubic phase. (10) Inverse hexagonal phase.

The molecularly templated synthesis approach is based on a surfactant-templated approach first reported by Mobil Oil Research.[16,17] Amphiphilic molecules, such as surfactant, lipids, copolymers, and proteins can self-assemble into a wide range of ordered structures.[18] These structures can also transform from one to another when the solution conditions, pH, temperature, or electrolyte concentrations are changed. The equilibrium structures are determined by the thermodynamics of the self-assembly process and the inter- and intra-aggregate forces. The major driving forces for the amphiphiles to form well defined aggregates are the hydrophobic attractions at the hydrocarbon-water interfaces and the hydrophilic ionic or steric repulsion between the head groups. A simple way to describe this kind of interaction is to use a critical geometric packing parameter, $v/a_o l_c$, where v is the volume of the hydrocarbon chains, a_o is the optimal head-group area, and l_c is the critical chain length. A small critical packing parameter (<0.5) favors the formation of a highly curved interface (spherical micelles and rod-like micelles), and a larger critical packing parameter (>0.5) favors the formation of flat interfaces (flexible bilayers and planar bilayers). A critical packing parameter larger than unity will produce inverse micelles. The surfactant geometry is related to the experimental conditions. Figure 1 is a simplified diagram of a surfactant-water-oil system. Depending on the solution compositions, spherical micelles, rod-like micelles, hexagonally ordered crystals, cubic crystals, lamellar phases, and inverse micelles, inverse micellar liquid crystals can be formed.

There may be several advantages in using the surfactant-templated approach to prepare low k dielectrics. When the ordered organic structures are used to template the synthesis of materials, we can precisely control and design the architecture of the ceramic materials on very fine scales. In particular, the ability to obtain uniform porosity may be beneficial in improving the mechanical strength of the highly porous structure. In addition, the potential to optimize the pore volume can lead to ultra-low k dieletric materials.

In this paper, we report on the synthesis, microstructure, and properties of mesoporous silica films prepared with two classes of surfactants. A range of experimental parameters, including the aging time of the deposition solution, the spinning rate, and the effect of solution treatments after surfactant pyrolysis, was investigated. The dielectric response, film quality, and processability using spin-coating technology are correlated to the experimental conditions.

EXPERIMENTAL APPROACH

We used the spin-coating technique to prepare the porous silica dielectrics. The process is schematically shown in Figure 2, which compares the surfactant-templated approach and the traditional approach for preparing porous silica dielectrics. Spin coating is a very fast and simple process that can be adapted by the semiconducting industry.

Spin coating of mesostructured films involves mixing silica precursors (tetraethyl orthosilicate [TEOS]) and surfactants, as well as co-solvents and catalysts, followed by rapid evaporation on silicon wafers. When the surfactant was mixed in the aqueous solution, the surfactant molecules were homogeneously dispersed. During the evaporation process, the elimination of the solvent caused the surfactant to form aggregated micellar structures in the film. Subsequent aging and heating served to condense the silicate into a continuous silica network around the micelles. The organic contents of the micelles were late removed by calcining above 400°C. Since the pore structure is templated by the surfactant micelles and has an extremely uniform size, the pore structure is stable during the drying and calcination process. The uniform pore channels may also help improve the mechanical strength of the film. In the traditional aero-gel or xero-gel approach, the silica network was caused by aggregation of fine colloidal particles.[19] The network structure is fractal and is subjected to significant shrinkage. The drying process needs to be tightly controlled to avoid the collapse of the pore structure. The pore network is not uniformly distributed, and the end material may have weak mechanical strength.

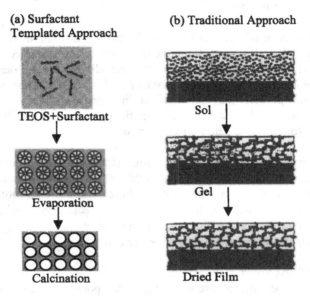

Figure 2. Comparison of surfactant-templated preparation and traditional aero-gel or xero-gel processing.

RESULTS AND DISCUSSION

The rapid evaporation process can be used to produce a wide range of morphologies, including spheres, fibers, and films (Figure 3).[20] The scanning electron microscopy (SEM) cross-sectional micrograph of the spin-coated film shows that the porous silica films prepared are very uniform in thickness, and the film texture is also very homogeneous.

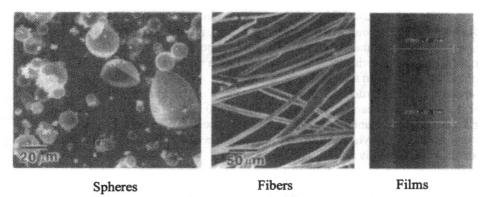

Spheres　　　　　Fibers　　　　　Films

Figure 3. Nanoporous spheres, fibers, and films prepared using rapid evaporation of surfactant-templated solutions.

However, the film preparation is a dynamic process with a series of concurrent physical and chemical reactions, including silicate hydrolysis and condensation, binding of the silicate species and the surfactant, and agglomeration of the surfactants. The nanostructure of the film and the quality depend not only on the surfactant used in the experiments, but also on the specific synthesis conditions. In the transmission electron (TEM) micrographs, completely ordered nanoporous films, partially ordered nanoporous films, and disordered nanoporous films have been observed. The porosity in these films can be as high as 65 volume percent. In both the ordered and disordered samples, the porosity and the pore size are very uniform, as verified by both TEM and other techniques, such as nitrogen adsorption.

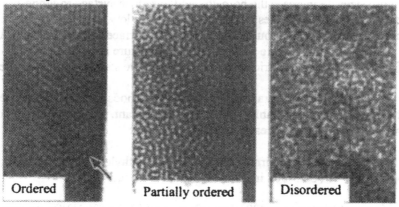

Ordered　　　　Partially ordered　　　Disordered

Figure 4. TEM micrographs of ordered, partially ordered, and disordered nanoporous silica films from the templated approach. The arrow points to a 3-nm pore.

Sol-Gel Commercialization and Applications

One of the critical parameters during spin coating is the surface roughness. Radial striation, a common defect in spun films, is undesirable for the current application. We found that radial striation can be significantly reduced by optimizing the experimental systems. Figure 5a shows the surface profile obtained from a film using a cationic surfactant. The surface roughness due to striation is on the order of 100 nm. However, if a nonionic surfactant is used, the surface roughness can be reduced to less than 10 nm. We believe in the case of the cationic surfactant that the striation is enhanced due to strong ionic binding between the hydrolyized silicate and the surfactant. The exact mechanism of striation formation is not fully understood yet.

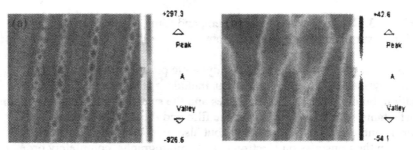

Figure 5. Striations formed with cationic surfactant (a), and nonionic surfactant (b). The height of the striation for the nonionic surfactant is within 10 nm.

To obtain stable low k dielectric films, we need to control the uniformity of the pore sizes, to increase the porosity, and to reduce surface roughness. However, a high porosity does not necessarily give a low dielectric constant. The dielectric properties of a porous silica film, with a surface area as high as 1000 m^2/g, are sensitive to moisture adsorption. The moisture adsorption would significantly raise the dielectric constant and affect the stability of the device.

We investigated several dehydroxylation methods to reduce moisture sensitivity and to obtain a stable, low-dielectric constant. One or more of the following dyhydroxylation treatments were followed:

(1) Immersion in a stirred solution of hexamethyldisilazane {HMDS, (CH3)3-Si-NH-Si-(CH3)30} in tolune for 20 to 24 hours.

(2) Heat treatment at 400°C in a flowing 2% H2-98% N2 forming gas.

(3) Heat treatment at 400°C in flowing argon.

(4) Spin-coating with pure HMDS at 2000 rpm.

As can be seen from Figure 6, all the dehydroxylation treatments were helpful in reducing the dielectric constant and stabilizing the film against aging in moist air. The film, without any treatment, gave a dielectric constant of 5 or higher. After treating with an inert gas, the dielectric constant was reduced to between 2.5 and 3.5, but still increased significantly when aged in air. Treating with HMDS in combination with an inert gas treatment provided a stable film with a dielectric constant between 1.7 and 2.2.

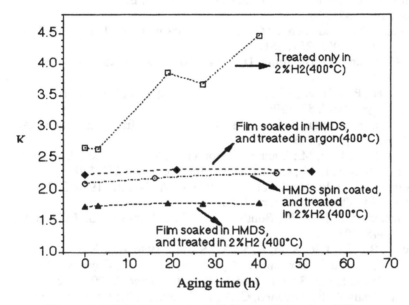

Figure 6. Dielectric constant of films subjected to different dehydroxylation treatment. A combination of HMDS and gas treatment gave a stable, low-dielectric constant between 1.7 and 2.2.[11]

In summary, we have demonstrated that the molecular-templated approach provides an effective route to prepare high-porosity (up to 65 volume %) nanoporous dielectric films with a uniform pore diameter smaller than 5 nm. The optimization of the experimental conditions, in combination of proper dyhydroxylation treatments, produced stable, low-dielectric constants between 1.8

and 2.2. The ability to increase and control the porosity and to optimize the materials parameters makes this approach attractive for next-generation ultralow dielectrics.

ACKNOWLEDGMENTS
Pacific Northwest National Laboratory is operated by Battelle for the U. S. Department of Energy under Contract DE-AC06-76RL0 1830.

REFERENCES

1. Smith, D. M.; Anderson, J.; Cho, C. C.; Gnade, B. E. MRS Soc. Symp. Proc., 1995, 371, 261.
2. Smith, D. M.; Anderson, J.; Cho, C. C.; Johnson, G. P.; Jeng, S. P. MRS Symp. Proc., 1995, 381, 261.
3. Jin, C.; Luttmer. J. D.; Smith, D. M.; Ramos, T. A. MRS Bull., 1997, 10, 39.
4. Jo. M.-H.; Park H.-H.; Kim, D. J.; Hyun, S.-H.; Choi, S.-Y.; Paik, J.-T. Appl. Phys., 1997, 82, 1299.
5. Yang, H.; Kuperman, A.; Coombs, N.; Mamiche-Afrara, S.; Ozin, G. A. Nature, 1996, 379, 703.
6. Aksay, I. A.; Trau, M.; Manne, S.; Honma, I.; Yao, N.; Zhou, L.; Fenter, P.; Eisenberger, P. M.; Gruner, S. M. Science, 1996, 273, 892.
7. Yang, H.; Coombs, N.; Sokolov, I.; Ozin, G. A. Nature, 1996, 381, 589.
8. Ogawa, M. Chem. Commun., 1996, 1149.
9. Bruinsma, P. J.; Hess, N.; Bontha, J. R.; Liu, J; Baskaran, S. MRS Bull., 1997, 445, 105.
10. Liu, J.; Bontha, J. R.; Kim, A. Y. MRS Bull., 1996, 431, 245.
11. Baskara, S.; Liu, J.; Domansky, K.; Kohler, N.; Li, X.; Coyle, C.; Fryxell, G. E.; Thevathasan, S.; Williford, R. E. Adv. Mater., 2000, 12, 291.
12. Lu, Y.; Ganguli, R.; Drewien, C. A.; Anderson, M. T.; Brinker, C. J.; Gong, W.; Guo, Y.; Soyez, H.; Dunn, B.; Huang, H. H.; Zink, J. I. Nature, 1997, 389, 364.
13. Zhao, D.; Yang, P.; Melosh, N.; Feng, J.; Chmelka, B. F.; Stucky, G. D. Adv. Mater., 1998, 10, 1380.
14. Evans, D. F.; Wennerstrom, H. The Colloidal Domain, Where Physics, Chemistry, Biology, and Technology Meet, VHC Publishers Inc, New York, 1994, Chapter 2.
15. Vinson, P. K.; Bellare, J. R.; Davis, H. T.; Miller, W. G.; Scriven, L. E. J. Coll. Int. Sci., 199, 142, 74.

16. Beck, J. S.; Vartuli, J. C.; Roth, W. J.; Leonowicz, M. E.; Kresge, C. T.; Schmitt, K. D.; Chu, C. T-W.; Olson, D. H.; Sheppard, E. W.; McCullen, S. B.; Higgins, J. B.; Schlenker, J. L. J. Am. Chem. Soc., 1992, 114, 10834.
17. Kresge, C. T.; Leonowicz, M. E.; Roth, W. J.; Vartuli, J. C.; Beck, J. S. Nature, 1992, 359, 710.
18. Israelachvili. J. Intermolecular & Surface Forces, 2nd Edition, Academic press, San Diego, 1991, Chapters 16, 17, and 18.
19. Brinker, C. J.; Schere, G. W. Sol-Gel Science, Academic Press, San Diego, 1990.
20. Bruisma, P. J.; Kim, A. Y.; Liu, J.; Baskaran, S. Chem. Mater., 1997, 9, 2507.

TITANIUM OXO-ORGANO CLUSTERS : PRECURSORS FOR THE PREPARATION OF NANOSTRUCTURED TITANIUM OXIDE BASED MATERIALS.

N. Steunou and C. Sanchez,
Chimie de la Matière Condensée,
UMR CNRS 7574,
Université P. et M. Curie,
4 place Jussieu, 75252 Paris, France

P. Florian
CNRS-CRPHT
1D, Av. Recherche
Scientifique
45071 Orléans cedex 2, France

S. Förster, C. Göltner, M. Antonietti,
Max Planck Institute für Kolloid- &
Grenzflächenforschung,
Am Mühlenberg
14476 Golm, Germany

ABSTRACT

Six titanium-oxo-organo clusters have been synthesized by adding titanium alkoxide to different complexing ligands such as carboxylic acids or ketones (acetone, diacetone alcohol, acetylacetone). In contrast to the classical sol-gel method, this approach does not imply the direct addition of water to the titanium alkoxide. The structure of theses complexes were determined by single crystal X-ray diffraction, ^{17}O and ^{13}C NMR spectroscopy. The nuclearity of these clusters increases with complexing ligand/Ti ratio and temperature.

Two different nanostructured "hybrid materials" based on titanium oxide were prepared by reacting the titanium-oxo clusters with an amphiphilic block copolymer PS-b-PMAA (PS=polystyrene block; PMMA=polymethacrylic acid block). These materials were characterized by TEM, SAXS, ^{17}O MAS NMR spectroscopy.

INTRODUCTION

The preparation of nanostructured materials is an important challenge of materials science. This field of chemistry is involved into the synthesis of advanced materials which are made with large assemblies of nanoparticles or metal based clusters. Size reduction of metals and semiconductors to the range of a few nanometers can lead to novel and peculiar properties. As an example, owing to their large active surface, the metal nanoparticles can exhibit interesting catalytic properties. Moreover, optical, electric or magnetic properties of a material are strongly dependent on the size and on the shape of the particle as well : optical transitions and spectral characteristics are directly correlated to the particle size of metal colloids (quantum size effects) [1]. Therefore the performance of the materials in applications requires the control over the size, the morphology and the surface structure, which is based on the appropriate control of the parameters that influence nucleation and growth. Such a control over the growth and morphology of the materials can be achieved by the use of organic templates which self-assemble into complex structures, patterning inorganic architectures built from inorganic precursors.

The synthesis of such materials through low temperature chemical routes allows a better control of their microstructure. Among the different processes involving low temperature, sol-gel processes are a way of making highly dispersed materials through the growth of metal oxo-polymers in a solvent [2-3]. They are based on hydrolysis-condensation reactions of molecular precursors such as metal alkoxides $M(OR)_4$ (M = Si, Ti...). However, as the metal alkoxide are particularly sensitive towards hydrolysis, the resulting oxo-polymers that constitute the sols and gels are generally polydisperse in size and composition.

An attractive solution is to use preformed metal-oxo clusters of well defined structure as precursors for making new hybrid organic-inorganic assemblies and new metal oxides. These complexes being already condensed compounds are less reactive towards nucleophilic species compared to metal alkoxides. Depending on the set of chosen experimental conditions these clusters can be considered as molecular building blocks or as reservoir of matter with a lower reactivity than that of the starting metal alkoxide.

The transition metal-oxo clusters are generally synthesized via substoichiometric hydrolysis ($H_2O/M<1$) of metal alkoxides [4] or of metal alkoxides complexed by some complexing ligands like β-diketones or carboxylic acids [5-6]. The complexing ligands are often used to modify the titanium alkoxides and to control their reactivity. However, it is well known that the carboxylates present a second advantage : by reacting the metal alkoxide with a carboxylic acid, water can be produced in-situ through esterification reactions with alkoxy ligands or with alcohol molecules [7]. As a consequence, this alternative strategy using in situ generation of water can be developed in order to synthesize new metal-oxo clusters [5,8,9,10]. Moreover, this approach avoids stability problems arising from local over-concentrations of water or the presence of residual water in the synthetic medium.

This paper deals first with the synthesis of titanium-oxo clusters which does not imply the direct addition of water to the titanium alkoxide. These syntheses are performed in the presence of acetic acid or different ketones. The second part of the paper describes the sol-gel synthesis of nanostructured hybrid materials from titanium-oxo clusters and amphiphilic block copolymers. The texturation of titanium-oxo clusters has already been performed through the assembly of these nano-building blocks with organic templates such as dendrimers [11] or amphiphilic block copolymers [12]. Amphiphilic block copolymers (ABCs) are consisting of hydrophilic and hydrophobic blocks, whose incompatibility can give rise to the self-assembly of block copolymers into particulate structures such as spherical and cylindrical micelles but also into lyotropic liquid crystalline phases. Moreover, these amphiphilic block copolymers have already been used as templates for the synthesis of metal nanoparticles with the desired shape and size [13] and for the synthesis of mesoporous oxides [14]. In this work, the hybrid materials are built from the association of titanium-oxo clusters and new amphiphilic block copolymers based on an alternance of polystyrene blocks and polymethacrylic acid blocks.

EXPERIMENTAL

Syntheses of titanium-oxo-organo clusters

All the reagents were purchased from Fluka. The complexes $[Ti_6O_4(OAc)_4(OPr^i)_{12}]$ (1), $[Ti_{12}O_{16}(OPr^i)_{16}]$ (2) and $[Ti_{17}O_{24}(OPr^i)_{20}]$ (3) were prepared as follows : titanium isopropoxide was added to acetic acid in a 1/1.2 molar ratio (for complexes (1) (2)) or in a 1/1.3 molar ratio (for complex (3)). The reaction mixtures were heated by reflux for two days at 80°C (for complexes (1) and (2)), or in a Parr Teflon-lined acid digestion bomb of 23-mL capacity at 150°C during 5 days (for complex (3)). At room temperature, crystals corresponding to complexes (1), (2) and (3) were grown from the reaction mixtures.

For the complex $[Ti_3O(OPr^i)_7(O_3C_9H_{15})]$ (4), titanium isopropoxide and dry propan2-one (acetone) in a 5/100 molar ratio were stirred in a closed vessel until a clear yellow solution was obtained. For the complex $[Ti_{11}O_{13}(OPr^i)_{18}]$ (5), titanium isopropoxide was added to dry 4-hydroxy-4-methyl-2-pentanone (diacetone alcohol) in a 1/1 molar ratio at room temperature. During stirring the resulting solution turned a red-brown color. From the reaction mixtures, crystals corresponding to complexes (4) and (5) were grown after an ageing period of 24H.

For the complex $[Ti_{16}O_{16}(OEt)_{32}]$ (6), titanium ethoxide was added to acetylacetone in a 1/0.6 molar ratio. The reaction mixture is heated in a Parr

Teflon-lined acid digestion bomb of 23-mL capacity at 150°C during 5 days. Crystals corresponding to complex (6) were obtained after heating.

Syntheses of the organic-inorganic "hybrid" materials

The PS-b-PMAA block copolymers chosen are consisting of one polystyrene block (PS) and one polymethacrylic acid block (PMAA) with equal chain lengths (105 monomer units). The synthesis of the PS-b-PMAA polymers which is mainly based on anionic polymerization was performed as described in earlier publication [15].

For the hybrid material PS-Ti-I, the block copolymer PS-b-PMAA was first dissolved in toluene at a concentration of 13.5 mg/mL. 25 mL of the PS-b-PMAA solution is mixed with 1 mL of a 120 mg/mL solution of the complex (6) in toluene (3 equivalent of complex (6) per PS-b-PMAA). The resulting solution is stirred for 1 H.

For the hybrid material PS-Ti-II, 25 mL of a 8.4 mg/mL PS-b-PMAA solution in toluene is mixed with 1 mL of a 77 mg/mL solution of the complex (6) in toluene (3 equivalent of complex (6) per PS-b-PMAA). After stirring the solution for a few minutes, the flask is evacuated in order to remove the toluene solvent. 25 mL of EtOH is then added to the flask and the resulting mixture is stirred for 1 H.

RESULTS AND DISCUSSION

Synthesis of the titanium oxo organo clusters in the presence of acetic acid or ketones.

The structure of the complexes (1) to (6) were determined previously by single crystal X-ray diffraction [4(a), 8, 10, 16, 17]

The synthesis of the titanium-oxo-alkoxide clusters $[Ti_6O_4(OAc)_4(OPr^i)_{12}]$ (1), $[Ti_{12}O_{16}(OPr^i)_{16}]$ (2) and $[Ti_{17}O_{24}(OPr^i)_{20}]$ (3) were performed by adding titanium isopropoxide to acetic acid. The molecular structures of these clusters are represented in Figure 1. These complexes can be distinguished by their nuclearity, their topology, the nature of the organic ligands and the coordinence of the metal centres.

Complex 1 is a hexameric cluster which contains titanium atoms with a distorted octahedral environment. These titanium centers are coordinated by acetato ligands and isopropoxy groups. Complex (2) is a highly condensed species (degree of condensation O/Ti= 1.33) which does not contain any carboxylate ligand [4(a)]. This cage-like cluster can be described as a hexameric cycle of six five-coordinate titanium centres, which is capped by two units of three sixfold coordinated titanium atoms. While the synthesis of the oxo-cluster $[Ti_{12}O_{16}(OPr^i)_{16}]$ in presence of acetic acid needs some heating (T>60°C), the hexameric cluster $[Ti_6O_4(OAc)_4(OPr^i)_{12}]$ can be obtained at room temperature [8]. Complex $[Ti_{17}O_{24}(OPr^i)_{20}]$ (3) is a cluster with 17 titanium atoms coordinated to oxo and isopropoxo ligands (Figure 1). It is a more condensed titanium cluster (O/Ti =1.41) with a metal-oxo core particularly compact compared to those of complexes (1) and (2). The metal-oxo framework is closely related to the Keggin structure (Figure 1). One feature of this cluster is the presence of titanium atoms with different environment : one is a tetrahedral centre, twelve are octahedral centres and four have a trigonal-bipyramidal environment.

For the complexes (1), (2) and (3), the source of condensation is water generated through esterification reactions between the acetic acid and isopropoxide ligands (or the cleaved alcohol). This method allows homogeneous hydrolysis and condensation of alkoxide precursors from an in-situ generated water. In contrast to complex (1), neither complex (2) nor complex (3) contain any carboxylate ligand. This indicates that, most of the carboxylic acid molecules have been transformed into isopropyl ester and water by increasing the temperature. Moreover, complex (3) was obtained with a larger amount of acetic acid (AcOH/Ti = 1.3/1) and at higher temperature (T=150°C) than complexes (1) and (2). Therefore, by increasing the amount of acetic acid and the temperature, a larger conversion of the esterification process is obtained and consequently a

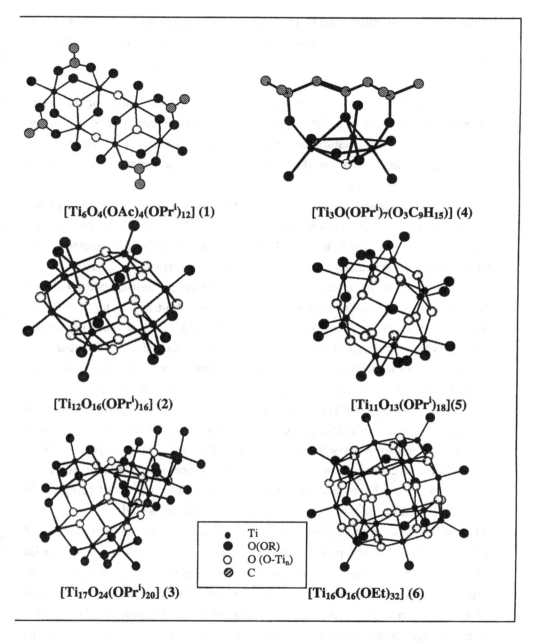

Figure 1 : Molecular structures of titanium-oxo-organo clusters (1)-(6). The carbon of alkoxy groups are omitted for clarity.

higher hydrolysis ratio h=H$_2$O/Ti [18]. As a consequence, these synthetic conditions lead to a more condensed titanium-oxo cluster like complex (3) [10,18].

This study shows that it is possible to achieve a better control of the water generated in-situ and thus a better control of the metal-oxo core size of the cluster by adjusting the temperature and the amount of acetic acid. Therefore by increasing the temperature and the amount of acetic acid, the metal-oxo cores of titanium clusters increase.

At room temperature, the complex [Ti$_3$O(OPr^j)$_7$(O$_3$C$_9$H$_{15}$)] (4) can be obtained in the presence of dry acetone while [Ti$_{11}$O$_{13}$(OPr^j)$_{18}$](5) is obtained in the presence of dry diacetone alcohol. These complexes were fully characterized by combining X-ray diffraction experiments and NMR spectroscopy [17,19].

The structure of complex (4) is based on a trinuclear unit capped by a new tridentate ligand (O$_3$C$_9$H$_{18}$ = 2,6-dimethylhept-3-en-2,4,6-triol). In this cluster, all the titanium atoms are six-coordinate. The complex (5) is a more condensed species whose structure is quite close to those of complex (2). Compared to complex (2), a vacancy is located in the cycle of pentacoordinated titanium atoms.

In the case of the complexes (4) and (5), the source of oxolation is provided by the ketones through ketolization and crotonization reactions which are catalyzed by the titanium alkoxide [19]. These reactions imply the condensation of carbonyl compounds and some dehydration steps. Actually, the tridentate ligand O$_3$C$_9$H$_{18}$ arises from the condensation of three molecules of acetone. Moreover, some ^{13}C NMR experiments have unambiguously shown that the synthesis of complex (5) is accompanied by the formation of mesityl oxide molecules [19].

The mesityl oxide molecules can be obtained through the dehydration of diacetone alcohol. These reactions may be catalyzed by the Lewis acid properties of titanium isopropoxide. In particular, it has already been proposed that the crotonization reactions may be accompanied by the hydroxylation of the titanium alkoxide which is therefore prone to condense [19]. This process constitutes a non-hydrolytic hydroxylation of the metallic center. A possible mechanism which accounts for the formation of the tridentate ligand and the complex (4) has been proposed [19].

The complex [$Ti_{16}O_{16}(OEt)_{32}$] (6) consists of two orthogonal blocks of eight TiO_6 octahedra. The formation of this complex is more difficult to explain. This complex was previously obtained by another sol-gel method based on the solvothermal treatment of titanium ethoxide [16]. In this work, the acetylacetone molecules act likely as oxolation sources for the synthesis of complex (6). However, the mechanism is not obvious as ethylacetate ester has been detected in the reaction bath through ^{17}O and ^{13}C NMR experiments [19]. The ester compound is likely formed through a retro-Claisen reaction from acetylacetone and was already evidenced in the case of a yittrium alkoxide derived complex [20].

NMR characterization of the titanium-oxo-organo clusters and stability study of these clusters face to nucleophilic species

All the complexes (1) to (6) were characterized by ^1H, ^{13}C, ^{17}O NMR spectroscopy in solution and by ^{13}C CP MAS in solid state [4(a), 8, 10, 19, 22]. Two of them namely complexes [$Ti_{12}O_{16}(OPr^i)_{16}$] (2) and [$Ti_{16}O_{16}(OEt)_{32}$] (6) were also investigated by ^{17}O MAS NMR experiments [21]. In particular, ^{17}O NMR spectroscopy is a very useful tool of characterization of the titanium-oxo clusters as the oxygen NMR parameters are sensitive to molecular structure and chemical environment. Oxygen-17 chemical shifts span over a range of 1500 ppm and along with quadrupolar coupling parameters provide detailed information on oxygen bonding, solvation, molecular structure. As a consequence, the ^{17}O NMR spectrum (number and intensity of the lines) of each cluster are fully consistent with the number and the nature of the oxo bridges O-Ti_n (n=2, 3, 4 or 5). The main drawback for the ^{17}O NMR spectroscopy in solution is the low natural abundance of the ^{17}O nuclei (0.037%) which can be overcome by hydrolyzing the titanium alkoxide precursors with ^{17}O-enriched water. In this work, as the titanium alkoxide is not directly hydrolyzed, the ^{17}O NMR spectra were recorded with extensive signal accumulation in order to get a good signal-to-noise ratio.

As an example, the ^{17}O MAS NMR spectrum of $[Ti_{16}O_{16}(OEt)_{32}](6)$ (Figure 2) displays one resonance in the region of the Ti_2-O oxygens (749 ppm), two in the region of the Ti_3-O oxygens (562, 554 ppm) and one in the region of Ti_4-O oxygens (382 ppm) [22, 23]. This spectrum is in good agreement with the nature and the number of oxo bridges in the cluster $[Ti_{16}O_{16}(OEt)_{32}]$.

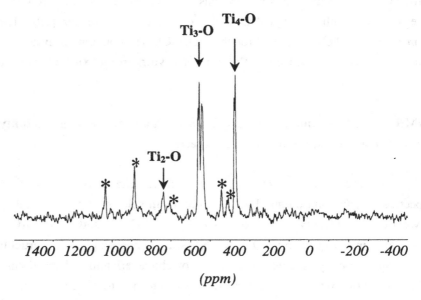

Figure.2 : ^{17}O MAS NMR spectrum of the cluster $[Ti_{16}O_{16}(OEt)_{32}]$ (v_{rot}(rotation speed) = 5 kHz; ^{17}O(Larmor Frequency) = 40.68 MHz; NS(number of scans) = 2860; LB(Line Broadening = 100 Hz) (* : spinning side bands).

The ^{17}O NMR data of the titanium oxo clusters (1)-(6) are given in Table 1.

Recording an ^{17}O NMR spectrum is particularly useful if one want to check the stability of the clusters in solution. Indeed, because of the presence of reactive alkoxy groups on the surface, the titanium-oxo clusters may exhibit a poor stability in presence of nucleophilic species. A systematic study on numerous titanium-oxo clusters leads to the conclusion that this stability is strongly dependent on the size of the clusters : as the size of the titanium oxo core increases, the stability of the corresponding cluster increases as well [22].

Table I ^{17}O NMR data for complexes (1)-(6) (solution NMR)

Complexes	δ (ppm)/ type /number of oxygen	reference
[Ti$_6$O$_4$(OAc)$_4$(OPr^j)$_{12}$] (1)	503 (1 ; Ti$_3$-O) 716 (1 ; Ti$_2$-O)	[8]
[Ti$_{12}$O$_{16}$(OPr^j)$_{16}$] (2)	517 (2 ; Ti$_3$-O) 522, 538, 563 (5 ; Ti$_3$-O) 815 (1 ; Ti$_2$-O)	[4(a), 8]
[Ti$_{17}$O$_{24}$(OPr^j)$_{20}$] (3)	436 (4 ; Ti$_4$-O) 533, 541 (16 ; Ti$_3$-O) 711 (4, Ti$_2$-O)	[10]
[Ti$_3$O(OPr^j)$_7$(O$_3$C$_9$H$_{15}$)] (4)	556 (1 ; Ti$_3$-O)	[19]
[Ti$_{11}$O$_{13}$(OPr^j)$_{18}$] (5)	515, 529, 536 (10 ; Ti$_3$-O) 711, 729, 770 (3 ; Ti$_2$-O)	[4(a), 19]
[Ti$_{16}$O$_{16}$(OEt)$_{32}$] (6)	382 (4 ; Ti$_4$-O) 562, 554 (8 ; Ti$_3$-O) 749 (4 ; Ti$_2$-O)	[22, 23]

Moreover, in the perspective of preparing some nanostructured hybrid materials, it is important to use the titanium-oxo cluster which presents the better stability towards nucleophilic species. The stability of three titanium oxo clusters of high nuclearity [Ti$_{12}$O$_{16}$(OPr^j)$_{16}$], [Ti$_{16}$O$_{16}$(OEt)$_{32}$], [Ti$_{18}$O$_{22}$(OBu^n)$_{32}$(acac)$_2$] towards different nucleophilic reagents (water, β-diketones, alcohol, carboxylic acid) was studied by NMR spectroscopy : ^{17}O solution NMR has been performed to assess the stability of the metal-oxo core while the ^{13}C solution NMR has been undertaken to study the lability of the organic surface ligands [12]. It has been shown that the stability of the titanium-oxo clusters towards nucleophilic species increase with the cluster nuclearity, the number of six-coordinate titanium atoms and the presence of bidentate ligands [12]. In particular, all the three clusters are stable in the presence of alcohol or polyols. In the presence of small amounts of water, the titanium-oxo cores can be conserved depending on the solvent used. Moreover, their stability towards strong complexing ligands (carboxylate, acetylacetonate) can be classified as follows : $[Ti_{12}O_{16}(OPr^j)_{16}]$ $<<[Ti_{16}O_{16}(OEt)_{32}]< [Ti_{18}O_{22}(OBu^n)_{32}(acac)_2]$. The first cluster is easily cleaved in the presence of small amounts of complexing ligands (RCOOH or acacH) while the two other are stable until the RCOOH/Ti and acacH/Ti ratios of 0.25 and 0.5 for $[Ti_{16}O_{16}(OEt)_{32}]$, and $[Ti_{18}O_{22}(OBu^n)_{32}(acac)_2]$ respectively are reached [12]. As the quite stable $[Ti_{16}O_{16}(OEt)_{32}]$ (6) cluster was obtained in a good yield, it was used as precursors for the preparation of nanostructured titanium-oxo based materials.

Synthesis and characterization of nanostructured titanium-oxo based materials.

For the present study, the PS-b-PMAA block copolymers chosen are consisting of one hydrophobic polystyrene block (PS) and one hydrophilic polymethacrylic acid block (PMAA). In contrast to the classical triblock copolymers poly(ethylene oxide)- poly(propylene oxide)- poly(ethylene oxide) (PEO-PPO-PEO, Pluronics), the solubility difference between the blocks is more pronounced in the case of PS-b-PMAA. One of the advantages of the block copolymers compared to low molecular weight analogues are the kinetic stability of the aggregated structures over a wide range of composition and temperature [24]. These templates have already been used for the preparation of mesoporous silica. As the order of the lyotropic ABC phase is not disrupted during the sol-gel synthesis of silica, the mesoporous materials are the exact replica of the parental mesophase and are therefore regular. Moreover they exhibit larger pore sizes (between 2 and 50 nm) and a higher wall thickness compared to ceramic oxides obtained from low molecular weight surfactants.

We have used different organic solvents that selectively dissolve one of the two blocks. In a selective solvent, the diblock copolymers form micelles with a core consisting of the insoluble block and a corona of the soluble block. Therefore, depending on the polarity of the solvent, different topologies may be formed from the self-assembly of block copolymers and titanium-oxo clusters.

Figure 3 shows transmission electron microscopy (TEM) micrographs of samples (PS-Ti-I) prepared by casting the solution on a carbon-covered microscopy grid. The picture represents hexagonal close packing of spherical diblock copolymers micelles with a diameter of about 30 nm. Some black dots with a size between 1 and 5 nm are located outside of the micelles cores in the polystyrene matrix or at the interface between the PMAA block and the PS block. These black regions are likely due to titanium-oxo based nanoparticles made of connected $[Ti_{16}O_{16}]$ oxo cores.

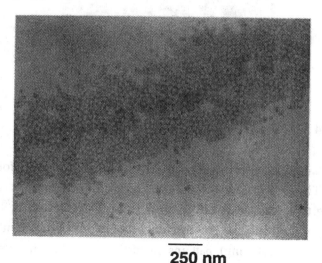

250 nm

Figure 3 : TEM micrograph of the "hybrid" material PS-Ti-I

Another morphology was observed when ethanol solvent was used instead of toluene. Ethanol is a polar solvent for the PMAA block. Therefore, micelles with a core of PS blocks and a corona of PMAA blocks are formed in EtOH. Figure 4 shows a TEM micrograph recorded on samples PS-Ti-II obtained by casting the solution. Block copolymer micelles 30 nm in diameter are present and no specific order is evidenced. Titanium-oxo based nanoparticles are located inside the micelles cores.

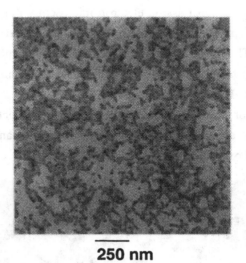

Figure 4 : TEM micrograph of the "hybrid" material PS-Ti-II

Small angle X-ray scattering (SAXS) experiments were performed on the samples PS-Ti-I and PS-Ti-II and the patterns exhibit one single peak corresponding to a Bragg distance of d=24.5 nm. This distance is in agreement with the size of the micelles. No diffraction peaks are observed in the WAXS (wide angle X-ray scattering) patterns indicating that the titanium-oxo based particles are amorphous. It is consistent with the fact that crystallization of titania from amorphous titanium-oxo aggregates requires usually higher temperature than 400°C.

In order to characterize the titanium-oxo based nanoparticles, ^{17}O ECHO MAS NMR experiments (spinning rates 10 kHz) were performed in solid state with ^{17}O-enriched samples. The ^{17}O NMR spectrum of the "hybrid" materials (PS-

Ti-I and PS-Ti-II) are identical. Only the spectrum of sample PS-Ti-I is represented in Figure 5. The spectra display two sets of resonances at 365 ppm and 540 ppm. These resonances can be assigned to the Ti_4-O and Ti_3-O oxygens respectively. These signals are quite close to those of the $[Ti_{16}O_{16}(OEt)_{32}]$ clusters spectrum (see Figure 2). Only the signal at 746 ppm in the spectrum of the clusters is missing in the spectra of the "hybrid" materials. This signal corresponding to the Ti_2-O oxygens exhibits a line width and a shape under the influence of a strong chemical shift anisotropy ($\Delta\sigma = 620$ ppm) [21]. It has been shown that upon the $[Ti_{16}O_{16}(OEt)_{32}]$ clusters connection these Ti_2-O signals are the most affected by the chemical modifications occurring at the shell of the $[Ti_{16}O_{16}]$ oxo core [11, 12]. In the "hybrid" materials, a wide chemical shifts distribution arising from a distribution of clusters with different orientations in the polymer matrix can explain a strong broadening of the signal corresponding to the Ti_2-O oxygens that consequently is not detected.

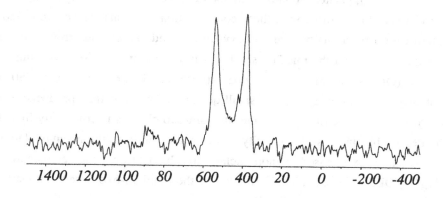

Figure 5 : ^{17}O ECHO MAS NMR spectrum of the "hybrid" material (PS-Ti-I) (ν_{rot}(rotation speed) = 10 kHz; ^{17}O(Larmor Frequency) = 54.25 MHz; NS(number of scans) = 512; LB(Line Broadening = 25 Hz).

Apparently, the metal-oxo cores of the $[Ti_{16}O_{16}(OEt)_{32}]$ clusters is preserved in the titanium-oxo based nanoparticles. Indeed, the ^{17}O NMR resonances are quite close to those of the parental titanium clusters and the line width of the signals ($\Delta v \sim 1600$ Hz) are weak compared to those of titanium oxo-polymers ($\Delta v \sim 3000-5000$ Hz). Moreover, it has already been demonstrated that the metal-oxo cores of the [Ti$_{16}$O$_{16}$(OEt)$_{32}$] clusters is quite stable in the presence of a small amount of water (16 equivalent of water per [Ti$_{16}$O$_{16}$(OEt)$_{32}$] cluster) [12, 22].

The location of the titanium-oxo based nanoparticles and consequently the nature of the interactions with the organic interface is more difficult to specify. On the one hand, the [Ti$_{16}$O$_{16}$(OEt)$_{32}$] clusters are soluble in non polar solvent as the metal oxo framework is capped by hydrophobic alkoxy groups. As a consequence, it can be expected that at the early stage of the process these clusters will be located in the hydrophobic region of the polymer matrix that is constituted by the polystyrene matrix. On the other hand, it is known that carboxylate ligands are strong complexing ligands. As a consequence, the titanium centers of the [Ti$_{16}$O$_{16}$(OEt)$_{32}$] clusters should be covalently bound to the carboxylate ligands of the PMAA blocks. Moreover, these complexation reactions are accompanied by esterification reactions between alkoxy ligands and carboxylate groups. Infrared spectra recorded on the micellar solution (in toluene or in EtOH) containing the [Ti$_{16}$O$_{16}$(OEt)$_{32}$] clusters display a small stretching vibration v(C=O) at 1750 cm^{-1} that is typical of ester species. A small quantity of water is then produced in situ suggesting that at least a part of the titanium-oxo clusters is covalently linked to the PMAA blocks. A small quantity of water released in solution can explain that the titanium-oxo based nanoparticles are built from the connection of [Ti$_{16}$O$_{16}$(OEt)$_{32}$] clusters. The formation of the titanium-oxo based nanoparticles proceeds probably through the hydrolysis-condensation of some alkoxy groups thereby forming bridges between [Ti$_{16}$O$_{16}$(OEt)$_{32}$] clusters.

On the basis of these experiments and our knowledge on the titanium-oxo clusters reactivity, it is possible to propose a mechanism for the formation of the "hybrid" material obtained in toluene (PS-Ti-I). The proposed mechanism is presented in Figure 6.

In toluene that selectively dissolves the PS block, the amphiphilic block copolymers PS-b-PMAA form spherical micelles with a core consisting of the insoluble block PMAA (Figure 6 (a)).

Once the titanium-oxo clusters are added to the micellar solution, their solubility in hydrophobic solvents leads to their incorporation in the PS corona (Figure 6 (b)). Some of them can diffuse into the amphiphilic interface between the PMAA block and the PS block. Indeed, the driving force of this diffusion is due to the great potentiality of carboxylate ligands to interact strongly with the titanium centers via covalent bonds (Figure 6 (c)). This interaction between titanium-oxo clusters and the hydrophilic block of the polymer may give rise to a decrease of the amphiphilic character of the polymer as it was previously reported for the texturation of the $[Ti_{16}O_{16}(OEt)_{32}]$ clusters by the triblock copolymer PEO-PPO-PEO [12]. As a consequence an inadequate folding of the template in the previous stages to the formation of the hybrid aggregates is responsible for the lack of any order in the hybrid materials [12].

In contrast to the polymer PEO-PPO-PEO, the block copolymer PS-b-PMAA presents a higher amphiphilic character. Moreover, as a large excess of the block copolymer is added (Ti/COOH ≈ 0.45), all the carboxylate groups of the PMAA blocks are not bound to the titanium atoms and the amphiphilic character of the polymer is conserved. As a consequence, the addition of the controlled amount of titanium-oxo clusters to the micellar solution does not disrupt the structures of the block copolymer micelles and an hexagonal mesophase is obtained upon evaporation of the solvent. Moreover, it is possible that the structure of the micelles and those of the hexagonal phase may be assisted by the addition of the titanium-oxo clusters. Indeed, the coordination of titanium atoms by carboxylate groups may be followed by some esterification reactions which promote the further hydrolysis and condensation of the titanium-oxo clusters (Figure 6 (d)).

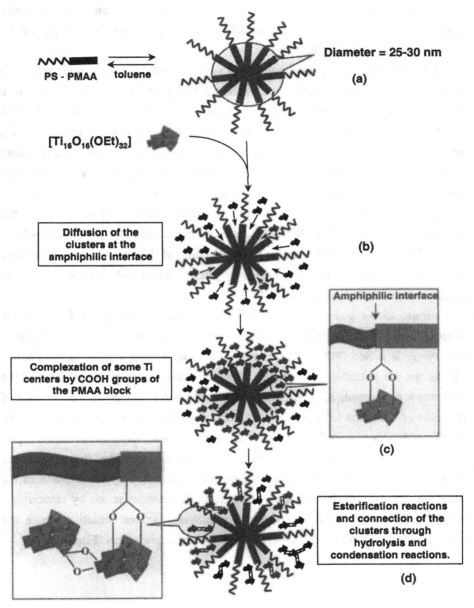

Figure 6 : Proposed mechanism for the formation of the hybrid PS-Ti-I.

The formation of an inorganic framework that is covalently linked to the polymer matrix enhances the stability of the aggregated structures.

For the second hybrid material (PS-Ti-II), as the block copolymer is dissolved in a polar solvent, reverse micelles are formed and the titanium-oxo clusters are located inside the cores of the micelles. Moreover, the solubilization of the polymer in the solvent EtOH may favour the esterification reactions and a larger amount of water may be released in the micellar solution. As a consequence, larger titanium oxide aggregates are formed.

CONCLUSION

Titanium oxide-block copolymer "hybrid" materials with different topologies can be obtained under the control of block copolymers micelles. The morphology of the materials is strongly dependent on the polarity of the solvent used to dissolve the block copolymer. In toluene, titanium-oxo based nanoparticles are located in the corona of micelles arranged in an hexagonal array while in ethanol, titanium-oxo based nanoparticles are incorporated in the micelles cores. Moreover, the titanium-oxo cores are preserved during the assembly, the complexation and the condensation processes. The stability of the titanium-oxo clusters demonstrates the great potential of these complexes as precursors for the controlled synthesis of titanium dioxide mesoporous materials.

References

[1] H. Weller, , "Colloidal semiconductor q-particles : chemistry in the transition region between solid state and molecules," *Angewandte chemie International Edition in English*, **32** 41 (1993).

[2] C.J. Brinker, G. Scherer, in *Sol-Gel Science, the Physics and Chemistry of Sol-Gel Processing*, Academic Press, San-Diego, CA, 1989.

[3] C. Sanchez, F. Ribot, "Design of hybrid organic-inorganic materials synthesized via sol-gel chemistry," *New Journal of Chemistry*, **18** 1007 (1994).

[4] (a) V.W. Day, T.A. Eberspacher, W.G. Klemperer, C.W. Park, "Dodecatitanates :A new family of Stable Polyoxotitanates,"*Journal of the American Chemical Society*, **115** 8469 (1993); (b) V.W. Day, T.A. Eberspacher, W.G. Klemperer, C.W. Park, F.S. Rosenberg, "Solution Structure Elucidation of Early-Transition-Metal Polyoxoalkoxides Using ^{17}O Nuclear Magnetic Resonance Spectroscopy,"*Journal of the American Chemical Society*, **113** 8190 (1991); (c) V.W. Day, T.A. Eberspacher, Y.Chen, J. Hao, W.G. Klemperer, "Low-nuclearity oxoalkoxides : the trititanates $[Ti_3O](OPr^i)_{10}$ and $[Ti_3O](OPr^i)_9(OMe)$,"*Inorganica Chimica Acta*, **229** 391 (1995).

[5] a) U. Schubert, E. Arpac, W. Glaubitt, A. Helmerich, C. Chau, "Primary Hydrolysis Products of Methacrylate-Modified Titanium and Zirconium Alkoxides," *Chemistry of Materials*, **4** 291 (1992); (b) S. Doeuff, Y. Dromzee, F. Taulelle, C. Sanchez, "Synthesis and Solid-and Liquid-State Characterization of a Hexameric Cluster of Titanium (IV),"*Inorganic Chemistry*, **28** 4439 (1989).

[6] P. Toledano, M. In, C. Sanchez, " Synthèse et structure du composé $[Ti_{18}(\mu_5\text{-}O)_2(\mu_4\text{-}O)_2(\mu_3\text{-}O)_{10}(\mu_2\text{-}O)_8(\mu_2\text{-}OBu^n)_{14}(OBu^n)_{12}(acac)_2]$,"*Comptes Rendus de l'Académie des Sciences Paris Série II*, **313** 1247 (1991).

[7] (a) C. Sanchez, F. Ribot, S. Doeuff, "Transition Metal Oxo Polymers via Sol-Gel Chemistry"; pp. 267-295 in *Inorganic and Organometallic Polymers with*

Special Properties, Edited by R.M. Laine. Kluwer, Dordrecht, Netherlands, 1992.

(b) A. Vioux, D. Leclercq, "Non Aqueous routes to Sol-Gel,"*Heterogeneous Chemistry Reviews*, **3** 65 (1996).

[8] N. Steunou, F. Robert, K. Boubekeur, F. Ribot, C. Sanchez, "Synthesis through an in situ esterification process and characterization of oxo isopropoxo titanium clusters,"*Inorganica Chimica Acta*, **279** 144 (1998).

[9] N. Steunou, C. Bonhomme, C. Sanchez, J. Vaissermann, L. G. Hubert-Pfalzgraf, " A Tetranuclear Niobium Oxo Acetate Complex. Synthesis, X-ray Crystal Structure, and Characterization by Solid-State and Liquid-State NMR Spectroscopy," *Inorganic Chemistry*, **37** 901 (1998).

[10] N. Steunou, G. Kickelbick, K. Boubekeur, C. Sanchez, "A new polyoxo-alkoxo titanium cluster of the Keggin family : synthesis and characterization by X-ray diffraction and NMR spectroscopy,"*Journal of the Chemical Society, Dalton Transactions*, 3653 (1999).

[11] G. J. A. A. Soler-Illia, L. Rozes, M.K. Boggiano, C. Sanchez, C-O. Turrin, A-M. Caminade, J-P. Majoral, "New Mesotextured Hydrid Materials made from Assemblies of Dendrimers and Titanium (IV) Oxo-Organo Clusters,"accepted in *Angewandte chemie*.

[12]G.J.A.A. Soler-Illia, E. Scolan, A. Louis, P.A. Albouy, C. Sanchez, "Design of Meso-Structured Titanium-Oxo based Hybrid Organic-Inorganic Networks,"accepted in *New Journal of Chemistry*.

[13] (a) V. Sankaran, J. Yue, R.E. Cohen, R.R. Schrock, R.J. Silbey, " Synthesis of Zinc Sulfide Clusters and Zinc Particles within Microphase-Separated Domains of Organometallic Block Copolymers," *Chemistry of Materials*, **5** [8] 1133 (1993); (b) J.P. Spatz, A. Roescher, M. Möller, "Gold Nanoparticles in Micellar Poly(styrene)-b-Poly(ethylene oxide) Films – Size and Interparticle Distance Control in Monoparticulate Films,"*Advanced Materials*, **8** [4]

337 (1996); (c) S. Klingelhöfer, W. Heitz, A. Greiner, S. Oestreich, S. Förster, M. Antonietti, "Preparation of Palladium Colloids in Block Copolymer Micelles and Their Use for the Catalysis of the Heck Reaction,"*Journal of the American Chemical Society*, **119** 10116 (1997).

[14] (a) D. Zhao, Q. Huo, J. Feng, B.F. Chmelka, G.D. Stucky, "Nonionic Triblock and Star Diblock Copolymer and Oligomeric Surfactant Syntheses of Highly Ordered, Hydrothermally Stable, Mesoporous Silica Structures,"*Journal of the American Chemical Society*, **120**, 6024 (1998) ;(b) E. Krämer, S. Förster, C. Göltner, M. Antonietti, " Synthesis of Nanoporous Silica with New Pore Morphologies by Templating the Assemblies of Ionic Block Copolymers,"*Langmuir*, **14**, 2027 (1998); (c) P. Yang, D. Zhao, B.F. Chemlka, G.D. Stucky, " Triblock-Copolymer-Directed Syntheses of Large-Pore Mesoporous Silica Fibers," *Chemistry of Materials*, **10** 2033 (1998) ; (d) C.G. Göltner, B. Berton, E. Krämer, M. Antonietti, "Nanoporous silica from amphiphilic block copolymer (ABC) aggregates:control over correlation and architecture of cylindrical pores,"*Chemical Communications*, 2287 (1998).

[15] C. Ramireddy, Z. Tuzar, K. Prochazka, S.E. Webber, P. Munk, " Styrene-tert-Butyl Methacrylate and Styrene-Methacrylic Acid Block Copolymers : Synthesis and Characterization,"*Macromolecules*, **25** 2541 (1992).

[16] R. Schmid, A. Mosset, J. Galy, "New compounds in the Chemistry of group 4 Transition-metal Alkoxides. Part 4. Synthesis and Molecular Structures of Two Polymorphs of [$Ti_{16}O_{16}(OEt)_{32}$] and Refinement of the Structure of [$Ti_7O_4(OEt)_{20}$],"*Journal of the Chemical Society, Dalton Transactions,* 1999 (1991).

[17]J.V. Barkley, J.S. Cannadine, I. Hannaford, M.M. Harding, A. Steiner, J. Tallon, R. Whyman, "Preparation and X-ray crystallographic characterization of

the trititanate [Ti$_3$O(μ-OPr^i)$_3$(OPr^i)$_4$\{Me$_2$C(O)CH=C(O)CH$_2$C(O)Me$_2$\}], a reaction product of [Ti(OPr^i)$_4$] and propan-2-one," *Chemical Communications* 1653 (1997).

[18] N. Steunou, R. Portal, C. Sanchez, "Carboxylic acids as an oxolation source for the synthesis of titanium oxo organo clusters,"*High Pressure Research*, in press.

[19] N. Steunou, F. Ribot, K. Boubekeur, J. Maquet, C. Sanchez, "ketones as an oxolation source for the synthesis of titanium-oxo-organo clusters,"*New Journal of Chemistry.*, **23** 1079 (1999).

[20] O. Poncelet, L. G. Hubert-Pfalzgraf, J-C. Daran, "Unexpected cleavage of acetylacetone induced by yttrium oxoisopropoxide: synthesis and molecular structure of Y$_2$(μ_2-OAc)$_2$(acac)$_4$(H$_2$O)$_2$," *Polyhedron*, **9** [10] 1305 (1990).

[21] E. Scolan, C. Magnenet, D. Massiot, C. Sanchez, "Surface and bulk characterisation of titanium-oxo clusters and nanosized titania particles through ^{17}O solid state NMR,"*Journal of Materials Chemistry*, 1999 **9** 2467.

[22] Y. W. Chen, W. G. Klemperer, C. W. Park, "Polynuclear Titanium Oxoalkoxides:Molecular Building Blocks for New Materials?," *Materials Research Society Symposium Proceedings,* **271** 57 (1992).

[23]J. Blanchard, S. Barboux-Doeuff, J. Maquet, C. Sanchez, "Investigation on hydrolysis-condensation reactions of titanium(IV) butoxide,"*New Journal of Chemistry*, **19** 929 (1995).

[24]S. Förster and M. Antonietti, " Amphiphilic Block Copolymers in Structure-Controlled Nanomaterial Hybrids," *Advanced materials*, **10** 195 (1998).

PROTON CONDUCTING SiO_2-P_2O_5-ZrO_2 SOL-GEL GLASSES

M. Aparicio and L.C. Klein
Rutgers, The State University of New Jersey
607 Taylor Rd.
Piscataway, NJ 08854-8065

ABSTRACT

Silicophosphate glasses have attracted much attention because of their potential as fast proton-conducting solids in energy generating applications, such as hydrogen fuel cells and proton exchange membrane fuel cells (PEMFC). Glasses in the SiO_2-P_2O_5-ZrO_2 system were prepared by the sol-gel method using tetraethylorthosilicate, triethyl phosphate, and zirconium n-propoxide. Monolithic, transparent gels were formed with 10 mole % ZrO_2, and 10, 20 and 30 mole % P_2O_5. The influences of the mole % P_2O_5, treatment temperature and water content on thermal behavior, water adsorption, and microstructure were studied. Heat treatment at 473 K leads to porous materials with high surface area (495 m^2/g) and a high capacity for water adsorption due to an abundance of M-OH bonds.

INTRODUCTION

Fuel cells are being developed for use in transportation, owing to their inherently high energy efficiency, compared to internal combustion engines. Fuel cells involving H_2 and O_2 (or air) are the only practical ones at the present time for mobile applications (1,2).

A major limitation of the current proton-exchange membrane fuel cells (PEMFC) is that the Pt anode electrocatalysts is poisoned by CO at the 5 to 10 ppm level. In addition, the Pt anode electrocatalyst is passivated by any unconverted fuel, such as methanol and ethanol, or other condensable intermediates, such as formaldehyde coming from the reformer. The current PEMFC also are complicated by the water management. For example, the proton conductivity of the PEMFC increases linearly with the water content of the membrane, with the highest conductivity corresponding to a fully hydrated membrane. While it is desirable to operate a fuel cell at a temperature above the boiling point of water from the standpoint of increased reaction kinetics and lower susceptibility to CO poisoning, the membranes lose conductivity due to drying. We propose to enhance the proton conductivity of ion-exchange membranes at elevated temperatures by impregnating the membrane with sol-gel processed silicophosphates. These gels are designed to provide a high concentration of protons by tying up the water in the pores and reducing its volatility, so that its vapor pressure is lower than that of free water (2-5).

Silicophosphate gels have been shown to be fast proton-conducting solids. The mobility of protons increases when the protons are strongly hydrogen-bonded. Compared with Si-OH, phosphate glasses are more efficient for high protonic conduction because the hydrogen ions are more strongly bound to the non-bridging oxygen. Also, the proton in the P-OH group is more strongly hydrogen-bonded with water molecules, resulting in an increase in the temperature necessary to remove the water from P-OH. At the same time, the presence of the silicate network is important because of its mechanical strength and chemical durability. The introduction of cations such as Zr^{4+} into silicophosphate gels results in improved chemical stability (6,7). The sol-gel process is suitable in this case because the low temperature allows infiltration into polymeric membranes, along with the preparation of microporous structures to improve the water retention through hydrogen-bonded protons on hydroxyl groups.

In this paper we report our results on gel formation in the SiO_2-P_2O_5-ZrO_2 system. Gels were heat treated to 473 K, and characterized to evaluate their water retention and porosity.

EXPERIMENTAL

SiO_2-P_2O_5-ZrO_2 glasses were prepared by the sol-gel method. Three molar compositions, $80SiO_2 \cdot 10P_2O_5 \cdot 10ZrO_2$ (8S1P1Z), $70SiO_2 \cdot 20P_2O_5 \cdot 10ZrO_2$ (7S2P1Z), and $60SiO_2 \cdot 30P_2O_5 \cdot 10ZrO_2$ (6S3P1Z), were prepared using tetraethyl orthosilicate [$Si(OC_2H_5)_4$, TEOS], triethyl phosphate [$PO(OC_2H_5)_3$, TEP], and zirconium-n-propoxide [$Zr(OC_3H_7)_4$, TPZr] as starting materials.

The sols were prepared by mixing two solutions. Solution A was prepared by mixing calculated amounts of TEOS, half the volume of propanol (solvent), TEP, HCl (to control the pH at ~2) and water [molar ratio of water/(TEOS+TEP)=2] at room temperature and stirred for 1 h. Solution B was prepared by mixing TPZr, the other half of the propanol and acetylacetone (molar ratio of acetylacetone/TPZr=1) at room temperature and stirred for 1 h. Both solutions were subsequently mixed together and stirred for another 1 h. The remaining amount of water was added drop by drop, and then the solution was stirred for 15 min. The sols have a concentration of 70 grams of solid per liter, and a final molar ratio of water/precursors=5.5. Ratios of 3 and 15 also were used with the 8S1P1Z composition to see the influence of the amount of water. The mixed sols were poured into centrifuge tubes that were sealed with plastic wrap, and placed at room temperature for gellation. One week after gel formation, several holes were punched in the warp to allow evaporation of the solvent. The samples were placed in an oven at 323 K. One month later, the samples were taken out of the drying oven and crushed. The dried gels were further heat treated at 423 and 473 K for one week.

Water adsorption was measured after 3 h in a humid atmosphere at room temperature by recording the weight change until the final weight was constant. The thermal behavior of gel powders was examined using a Perkin-Elmer TGA6 and DTA7 models, with a typical heating rate of 10 K/min in nitrogen. Specific surface area and pore volume measurements were made from the nitrogen gas adsorption isotherms (Coulter SA 3100). Fourier transform infrared (FTIR) spectra were recorded using a Perkin-Elmer 1720X in the frequency range 4000-400 cm^{-1}. FTIR measurements were made at room temperature using KBr pellets containing 0.5 wt % of powdered glass.

RESULTS AND DISCUSSION

All the gels dried at 323 K remained monolithic and transparent. The gellation time for the molar ratio of water/precursors=5.5 was 2 days for all three compositions. When this ratio was decreased to 3 the gellation time increased to 5, and when the ratio was increased to 15 the time decreased to 1 day.

Figure 1 shows TGA curves for 8S1P1Z composition with different water/precursors ratios dried at 323 K. The derivative curves (not shown) indicate one maximum centered at 593-603 K, that can be attributed to the removal of organic material and some water associated with the condensation of M-OH groups. At low temperature (~373 K), the weight loss is small because the long drying at 323 K has removed almost all residual water. The total weight loss increases for lower ratios of water to precursors because of a higher residual M-OR content. The residual M-OR are assumed to be P-OEt because the hydrolysis of TEOS, and of course TPZr, are reported to be much faster than that of TEP in alcoholic solvent (8-10).

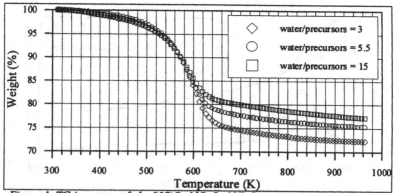

Figure 1. TGA curves of the $80SiO_2 \cdot 10P_2O_5 \cdot 10ZrO_2$ composition prepared with different water/precursors ratios and dried at 323 K.

The weight loss during the 473 K treatment and the weight gain due to the water adsorption process after the 423 and 473 K treatments are listed in Table I.

Table I. Weight variation after heat treatment and water adsorption

Molar composition	H_2O/ Precursors	Weight loss at 473 K (%)	Water adsorption (%)	
			After 423 K	After 473 K
8S1P1Z	3.0	21	3.7	12
	5.5	20	4.9	12
	15	17	4.1	10
7S2P1Z	5.5	32	4.8	10
6S3P1Z	5.5	41	3.9	10

The weight loss during the 473 K treatment increases with the percentage of P_2O_5 because the amount of residual organic groups, attached mainly to the phosphorus atom, also increases (11). The weight loss decreases when the initial water content is higher for the 8S1P1Z composition in good agreement with the TGA results. When the drying temperature increases from 423 to 473 K, the weight gain due to the water adsorption process rises almost one order of magnitude. The 423 K treatment only removed a small amount of organic material leaving many pores not available for the water adsorption. In contrast, the 473 K treatment removed higher percentage of residual organic material. This temperature is the lower limit for the elimination of vicinal surface OH groups to form M-O-M bonds and water.

Figure 2 presents TGA curves of 6S3P1Z samples after 423 and 473 K treatments, followed by the water adsorption process. The 423 K curve shows two weight losses with maximum rates at ~ 393 K due to water desorption and 600 K due to removal of organic material. In contrast, the 473 K curve shows only the weight loss at low temperature, indicating that treatment at 473 K has eliminated almost all organic material. In this case, the weight loss at low temperature is larger than that with the 423 K treatment, because of the higher amount of adsorbed water.

The remaining compositions have similar behavior. The weight loss decreases when the P_2O_5 percentage increases. This behavior is probably related to the small amount of M-OH groups because of the slower hydrolysis rate of TEP.

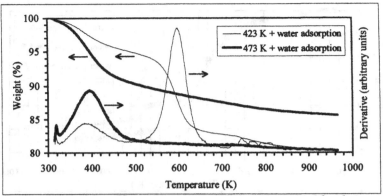

Figure 2. TGA curves of the $60SiO_2 \cdot 30P_2O_5 \cdot 10ZrO_2$ composition after drying at 423 and 473 K, and water adsorption process.

Figure 3 shows the DTA thermographs in nitrogen flow of the 8S1P1Z composition with a ratio water/precursors=5.5 at different temperatures.

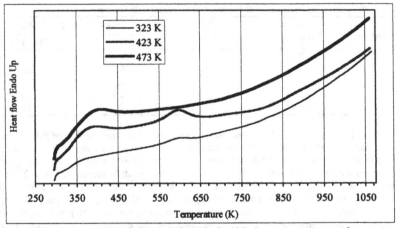

Figure 3. DTA curves of the $80SiO_2 \cdot 10P_2O_5 \cdot 10ZrO_2$ composition with a ratio water/precursors=5.5 at different temperatures.

The temperature ranges in the curves correspond to the changes observed in the TGA tests. There is an endothermic peak at low temperature between room temperature and 470 K due to the water desorption (12,13). In the sample without water adsorption (323 K), the peak is negligible. The maximum of the sample heated previously at 473 K is located at a slightly higher temperature than that at 423 K. This displacement coincides with a similar trend in the weight loss measurement. There are other endothermic peaks at higher temperature (between 530 and 650 K) corresponding to the decomposition of organic materials, mainly EtOH from the phosphorus precursor (13). The 473 K heated samples do not show this peak because the removal of organics is more complete.

The data from surface measurement area support the results of thermal analysis. Figure 4 shows the adsorption-desorption isotherms of the 8S1P1Z composition with water/precursors=3 heated at different temperatures.

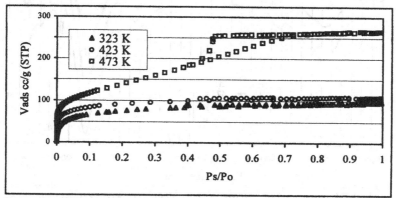

Figure 4. Adsorption-desorption isotherms of the $80SiO_2 \cdot 10P_2O_5 \cdot 10ZrO_2$ samples prepared with a water/precursors=3 and dried at different temperatures.

The samples heated at 323 and 423 K show Type I isotherms. This type of isotherm is characteristic of microporous materials (presence of pores with diameter less than 2 nm). The slight increase of the micropore volume with the treatment temperature, and increase in the values obtained for BET surface areas from 275 m^2/g at 323 K to 330 m^2/g at 423 K treatment, are due to the increase of the removal of water and organics. When the temperature rises to 473 K, the shape of the curve changes. In this case, the isotherm is Type IV indicating the presence of mesoporosity (2-50 nm). In this case, the presence of a high value of volume adsorbed at low P_s/P_o and a high value of the BET surface area (495 m^2/g) is an indication of the presence of microporosity. The 473 K treatment has been effective in removal of residual water and organics, creating a porous material capable of water retention. At a higher temperature, a decrease in the specific surface area is expected due to the condensation reaction between hydroxyl groups.

The room temperature FTIR transmission spectra of the heat-treated 8S1P1Z gel samples at different temperatures are shown in Figure 5. Assignments of the different bands are based on literature values (12-16). The highest band at 1060 cm^{-1} is assigned to the Si-O and P-O stretching vibrations associated with polymerized sheet structures. This band, with the one that appears in the 3000-3700 cm^{-1} region assigned to the stretching modes of OH groups, indicates that nonbridging oxygens are linked to hydrogen atoms. The band at 1635 cm^{-1}, generally assigned to the deformation modes of OH groups and adsorbed water molecules, support this point. These two bands do not change with the temperature of the treatment because the OH groups are stable at least up to 473 K. The presence of Si-OH bonds also was confirmed by the band at 590 cm^{-1}. The small peak at 2930 cm^{-1} can be attributed to the hydrogen-bonded OH groups. A shoulder at ~ 1160 cm^{-1} is assigned to vibration modes of P-O bonds. There is a shoulder at ~ 1260 cm^{-1} that has been attributed to P=O stretching of the phosphate units. This band usually appears in the regions 1320-1200 cm^{-1}, although the specific vibration frequency in any particular compound is determined by several factors, the most important being the total electronegativity of the groups attached to the P atom.

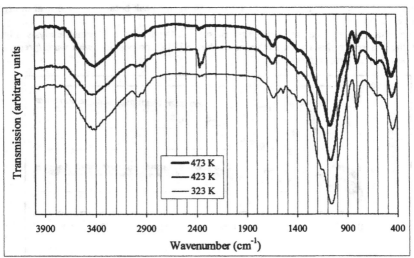

Figure 5. FTIR transmission spectra of the $80SiO_2 \cdot 10P_2O_5 \cdot 10ZrO_2$ composition prepared with a water/precursors=5.5 and dried at different temperatures.

There is a shoulder at about 970 cm^{-1} that can be ascribed to Si-O bends of non-bridging oxygen, although different bands in this same region are also characteristic of vibrations of the PO_4^{3-} monomer species and Si-O-P linkages. The peaks at 800 and 450 cm^{-1} can be associated mainly to bending modes of Si-O-Si and the ring structure of SiO_4 tetrahedra. The double peak at around 2350 cm^{-1} is assigned to P-OH stretching vibrations. The small peaks at 2965 and 1540 cm^{-1} disappear when the temperature increases and are thought to be due to the free water present in the gel. Finally, the peak at ~1390 cm^{-1} can be assigned to the bridging OH groups in the $Zr(OH)_4$ units, which decrease with increasing temperature to produce Zr-O-Zr bonds.

The FTIR transmission spectra of the 8S1P1Z composition with different water contents do not show significant differences. On the other hand, the increase of P_2O_5 percentage produces small variations, raising slightly the intensity and/or width of different bands related to the phosphorus atom, for example, the shoulder at 1260 cm^{-1}, the shoulder at 970 cm^{-1} and the double peak at 2350 cm^{-1}. It is reasonable to say that the structure of these heated gels after 473 K treatment consists of a siloxane framework containing silanols groups and small chains of metaphosphoric acid with one oxygen being linked by a double bond to phosphorus (P=O).

CONCLUSIONS

Monolithic and transparent gels with $xSiO_2 \cdot (90-x)P_2O_5 \cdot 10ZrO_2$ compositions have been prepared. The weight gain due to water adsorption more than doubles from ~4-5 % to 10-12 % when the previous treatment temperature increases from 423 to 473 K. The TGA-DTA results show that the main difference between the two treatments is the effectiveness of the removal of organics, mainly OEt groups attached to the phosphorus atom. This behavior increases the surface area from 330 m^2/g at 423 K to 495 m^2/g at 473 K. In conclusion, the 473 K temperature leads to a porous material consisting of a siloxane framework containing silanols and metaphosphoric acid groups. This structure is capable of water retention by hydrogen bonding to the abundant surface OH groups.

ACKNOWLEDGEMENTS

M. Aparicio wishes to thank the Spanish Ministry of Education and Culture for his financial support (grant PF99 0001828914). The authors thank Dr. T. Hao, Dr. C. A. Paredes and the rest of the Rutgers sol-gel research group for their encouragement and assistance.

REFERENCES

[1] G. Hoogers, "Fuel Cells: Power for the Future," *Physics World*, **11** [8] 31-6 (1998).
[2] M. Jacoby, "Fuel Cells Heading for Sale," *C&EN*, **14** June 31-7 (1999).
[3] M. Bhamidipati, E. Lazaro, F. Lyons and R. S. Morris, "Novel Proton Exchange Membrane for High Temperature Fuel Cells," *Material Research Society Symposium Proceeding*, **496** 217-22 (1998).
[4] M. Wakizoe, O. A. Velev and S. Srinivasan, "Analysis of Proton Exchange Membrane Fuel Cell Performance with Alternate Membranes," *Electrochimica Acta*, **40** [3] 335-44 (1995).
[5] S. Malhotra and R. Datta, "Membrane-Supported Nonvolatile Acidic Electrolytes Allow Higher Temperature Operation of Proton-Exchange membrane Fuel Cells," *Journal of Electrochemical Society*, **144** [2] L23-6 (1997).
[6] M. Nogami, K. Miyamura and Y. Abe, "Fast Protonic Conductors of Water-Containing P_2O_5-ZrO_2-SiO_2 Glasses," *Journal of Electrochemical Society*, **144** [6] 2175-8 (1997).
[7] M. Nogami, R. Nagao, W. Cong and Y. Abe, "Role of Water on Fast Proton Conduction in Sol-Gel Glasses," *Journal of Sol-Gel Science and Technology*, **13** 933-6 (1998).
[8] F. Tian, L. Pan, X. Wu and F. Wu, "The NMR Studies of the P_2O_5-SiO_2 Sol and Gel Chemistry," *Journal of Non-Crystalline Solids*, **104** 129-34 (1988).
[9] S. P. Szu, L. C. Klein and M. Greenblatt, "Effect of Precursors on the Structure of Phosphosilicate Gels: ^{29}Si and ^{31}P MAS-NMR Study," *Journal of Non-Crystalline Solids*, **143** 21-30 (1992).
[10] J. C. Schrotter, A. Cardenas, M. Smaihi and N. Hovnanian, "Silicon and Phosphorus Alkoxide Mixture: Sol-Gel Study by Spectroscopic Techniques," *Journal of Sol-Gel Science and Technology*, **4** 195-204 (1995).
[11] Z. Cao and B. I. Lee, "Sol-Gel Synthesis of Phosphate Ceramic Composites II," *Journal of Materials Research*, **13** [6] 1553-9 (1998).
[12] K. J. Rao, N. Baskaran, P. A. Ramakrishnan, B. G. Ravi and A. Karthikeyan, "Structural and Lithium Ion Transport Studies in Sol-Gel Prepared Lithium Silicophosphate Glasses," *Chemistry Materials*, **10** 3109-23 (1998).
[13] M. D'Apuzzo, A. Aronne, S. Esposito and P. Pernice, "Sol-Gel Synthesis of Humidity-Sensitive P_2O_5-SiO_2 Amorphous Films," *Journal of Sol-Gel Science and Technology*, **17** 247-54 (2000).
[14] E. A. Hayri and M. Greenblatt, "The Preparation and Ionic Conductivity of Sol-Gels in the Li_2O-P_2O_5-SiO_2 System," *Journal of Non-Crystalline Solids*, **94** 387-401 (1987).
[15] E. A. Hayri, M. Greenblatt, M. T. Tsai and P. P. Tsai, "Ionic Conductivity in the M_2O-P_2O_5-SiO_2 (M=H, Li, Na, K) System Prepared by Sol-Gel Methods," *Solid State Ionics*, **37** 233-7 (1990).
[16] Y. S. Kim and R. E. Tressler, "Microstructural Evolution of Sol-Gel Derived Phosphosilicate Gel with Heat Treatment," *Journal of Materials Science*, **29** 2531-5 (1994).

THE STUDY OF HYDROLYSIS AND CONDENSATION OF MIXED SOLUTION FROM $Si(OC_2H_5)_4$ AND $Zr(O-nC_3H_7)_4$

Dae-Yong Shin and Sang-Mok Han
C.A.C.P., Kangwon Univ., Chunchon,
Kangwon Do, R.O.K. 200-701

Kyung-Nam Kim
Dept. Mater. Eng., Samchuk Univ., Samchuk
Kangwon Do, R.O.K. 245-711

Wi-Soo Kang
Dept. Agri. Machi. Eng., Kangwon Univ.,
Chunchon, Kangwon Do, R.O.K. 200-701

Sang-Kyu Kang
KINITI, Chungryangri Dong 206-9,
Dongdaemun Gu, Seoul, R.O.K. 130-742

ABSTRACT

Monolithic gels of $ZrO_2 \cdot SiO_2$ containing up to 30 mol% of ZrO_2 were prepared from $Zr(O-nC_3H_7)_4$ and $Si(OC_2H_5)_4$. The reaction of hydrolysis and condensation of mixed solution of $Si(OC_2H_5)_4$ and $Zr(O-nC_3H_7)_4$, structural change from sol to gel were analyzed by GC and IR spectra.

INTRODUCTION

Sol-gel process based on the hydrolysis and condensation of metal alkoxide was still not fully understood but pose a number of questions which required further investigation. It is necessary to investigate in detail the reaction of hydrolysis and condensation of $Si(OC_2H_5)_4$ and $Zr(OC_3H_7)_4$ in order to prepare $ZrO_2 \cdot SiO_2$ glass with desired properties. The reactivity of $Zr(OC_3H_7)_4$ toward hydrolysis being higher than that of $Si(OC_2H_5)_4$[1], therefore a mixture of $Si(OC_2H_5)_4$ and $Zr(OC_3H_7)_4$ precipitate during hydrolysis instead of forming a homogeneous gel[2]. Determining the hydrolysis behavior of $Si(OC_2H_5)_4$ and $Zr(OC_3H_7)_4$ with various conditions can be useful for preparation of multicomponent gels.

In the present work, the hydrolysis and condensation of the mixed solution of $Si(OC_2H_5)_4$ and $Zr(O-nC_3H_7)_4$, structural change from sol to gel and gel-glass transition were analyzed by GC and IR spectra.

EXPERIMENTAL PROCEDURE

$Si(OC_2H_5)_4$ was at first partially hydrolyzed by dropping it into the mixture of 0.3 mol HCl solution, 1 mol of H_2O and C_2H_5OH to TEOS. After stirring this solution for 2 hr, $Zr(O-nC_3H_7)_4$, $H_2O(R=1\sim 8$ mol) and 0.3 mol of HCl were added drop by drop, followed by stirring for 2 hr. Gelation was left to occur at room temperature in teflon containers. The gels were dried at 110℃ for few days in an oven. The heat treatment was carried out in an electrical furnace in air, at a heating rate of 0.5℃/min up to 300℃, 500℃, 700℃ and 800℃ and maintaining the gels at each temperature for 10 hr.

The changes in the amount of alcohol and H_2O during partial hydrolysis of TEOS, mixed solution of TEOS and $Zr(O-nC_3H_7)_4$) were investigated using GC. The structural changes of hydrolysis solution and gels were examined by IR spectroscopy.

RESULTS AND DISCUSSION

The hydrolysis reactions of TEOS may be represented as follows[3],

$Si(OR)_4 + H_2O \rightarrow (RO)_3Si(OH) + ROH$ (1)

$Si(OR)_4 + (HO)Si(OR)_3 \rightarrow (RO)_3Si-O-Si(OR)_3 + ROH$ (2)

$(RO)_3Si(OH) + (HO)Si(OR)_3 \rightarrow (RO)_3Si-O-Si(OR)_3 + H_2O$ (3)

where R is C_2H_5.

Table 1 shows the quantity in the amount of respective compounds with reaction time. The amount of H_2O and C_2H_5OH varied with the reaction time elapsed, but the amount of MEK remained unchanged. The total generation amount of C_2H_5OH during partial hydrolysis of TEOS was obtained by both eq.(1) and (2)[4]. Theoretically, the hydrolysis of TEOS in 4 mol of H_2O lead to the formation of $Si(OH)_4$ and 4 mol of C_2H_5OH. In this study, the hydrolysis of TEOS was carried out in 1 mol of H_2O, partly hydrolyzed condensed silicates were produced in eq.(1)[5]. The formation of the polysilicates was the results of a complex sequence of reaction in eq.(2) and (3). In this study, the amount of H_2O was decreased with reaction time, therefore Si-O-Si bonding was completed by not eq.(3) but (2). It should be suggested that about a 3/4 of the unreacted TEOS still remained in solution.

$Zr(O-nC_3H_7)_4$ reacts with the end or the side silanol groups of the siloxane polymer to produce the copolymerized products as the following reactions[6].

$(RO)_3Si(OH) + (R'O)_4Zr \rightarrow (RO)_3Si-O-Zr(OR')_3 + R'OH$ (4)

$Zr(OR')_4 + H_2O \rightarrow (R'O)_3Zr(OH) + R'OH$ (5)

$(R'O)_3Zr(OH) + (HO)Zr(OR')_3 \rightarrow (R'O)_3Zr-O-Zr(OR')_3 + H_2O$ (6)

$(RO)_3Si(OH) + (HO)Zr(OR')_3 \rightarrow (RO)_3Si-O-Zr(OR')_3 + H_2O$ (7)

$Si(OR)_4 + (HO)Zr(OR')_3 \rightarrow (RO)_3Si-O-Zr(OR')_3 + ROH$ (8)

where R' is C_3H_7OH. Addition of $Zr(O-nC_3H_7)_4$, $Zr(OR)_4$ or $Zr(OH)_n(OR')$ species in eq.(5) reacts with silanol groups and polycondensation take place in eq.(4), (7) and (8). Adding $Zr(O-nC_3H_7)_4$ in partial hydrolysis solution of TEOS, the reactions in eq.(4), (5), (7) and (8) occurred simultaneously. The polymeric solution was clear and gelation occurred in a few days continued polymerization. It was suggested that $Zr(O-nC_3H_7)_4$ should be reacted with the hydrolyzed TEOS. $Zr(O-nC_3H_7)_4$ caused the self polymerization according to eq.(6) to result in the precipitation of zirconium hydroxide[6]. In this study, the reaction of eq.(6) was not proceeded because of no precipitation of zirconium hydroxide.

Table I. The quantity in the amount of compounds with reaction time

Step	Reaction time (min)	H_2O (mol)	C_2H_5OH (mol)	C_3H_7OH (mol)	MEK (mol)
1'st	0	1.04	0.97		4.02
	30	0.18	1.73		4.00
	60	0.08	1.84		4.01
	90	0.03	1.87		4.01
	120	0.02	1.88		4.02
2'nd	130	0.07	1.94	0.40	4.00
	150	3.40	3.32	2.90	4.01
	180	1.80	4.43	3.30	4.01
	210	0.03	4.93	3.90	4.02
	240	0.03	4.92	3.92	4.01

1'st : Partial hydrolysis solution of TEOS
2'nd : Mixed solution of TEOS and $Zr(O-nC_3H_7)_4$

Fig.1 shows the IR spectra of the C_2H_5OH(a), TEOS(b) and partial hydrolysis solution of TEOS(c). The absorption band of Si-O around 900 cm^{-1} and the characteristic band of C_2H_5OH around 1050 cm^{-1} were presented in (c). The characteristic band of TEOS around 790 cm^{-1} was decreased considerably, but it was still presented. It was suggested that the hydrolysis of TEOS was not completed.

Fig.2 shows the IR spectra of $30ZrO_2 \cdot 70SiO_2$ gels with aging time. The absorption bands of C-H around 2970, 1460 and 1380 cm^{-1} were still presented in the gels and these bands became weakened with aging time. It was suggested that the residual organic compounds were gradually evaporated and the condensation of the gels during aging was progressed. The absorption bands around 1200, 1080, 800 and 460 cm^{-1} were all characterized by the Si-O bond[7]. The fact that these absorption bands appear in the gels heated at low temperature confirms the formation of the Si-O anionic network structure in the gels. These bands were gradually sharpness with aging time.

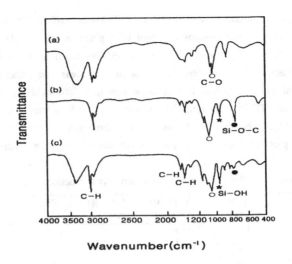

Fig.1. IR spectra of ethanol (a), TEOS(b) and partial hydrolysis solution of TEOS(c).

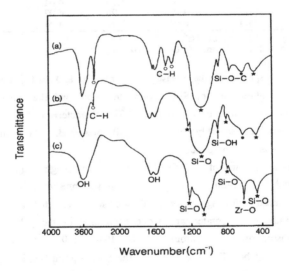

Fig.2. IR spectra of $30ZrO_2 \cdot 70SiO_2$ gels with aging time. 1 hr(a), 7 days(b) and 14 days(c).

Fig.3 shows the IR spectra of $30ZrO_2 \cdot 70SiO_2$ gels with heating temperature. The absorption band around 900 cm^{-1}, the intensity of which decreased with increasing temperature, was associated with the vibration of Si-O$^-$ bonds. It was indicated that this band could be associated with Si-O$^-$ bonded to a proton, none with Si-O$^-$ bonded to Zr^{4+} ion[8]. The absorption band which could be associated with the Zr-O$^-$ bond was seen around 600 cm^{-1}. This absorption band can be associated with the ZrO_8 groups in which the Zr^{4+} ion does not enter in the total anionic frame work structure of glass[7]. On heating the gels, the intensity of the band around 800 cm^{-1} increased instead of decreasing as for the band around 900 cm^{-1}, which indicates the formation of Si-O-Si bond by the dehydration-condensation reaction.

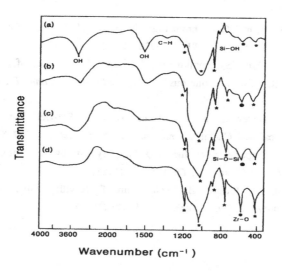

Fig.3. IR spectra of $30ZrO_2 \cdot 70SiO_2$ gels heated at 300℃(a), 500℃(b), 700℃(c) and 800℃(d).

CONCLUSION

The amount of H_2O and C_2H_5OH were changed with reaction time, Si-O-Si bonding and the generation of C_2H_5OH were obtained both by hydrolysis and liberated by condensation. $Zr(O-nC_3H_7)_4$ reacted with the partly hydrolyzed of TEOS, the clear polymeric solution was obtained. The hydrolysis was not completed because characteristic bands of organic compounds were presented in IR spectra of partial hydrolysis solution of TEOS and ZrO_2.

SiO_2 dried gel. The absorption bands due to C-H bonds were diminished with aging time and the absorption bands of Si-O bonds appeared in dried gels heated at low temperature confirms the formation of the Si-O and Zr-O anionic network structure in the gels.

REFERENCES

[1] R.C. Mehrota, "Synthesis and Reaction of Metal Alkoxides," *J. Non-Cryst. Sol.*, **100** 1-15 (1985).

[2] M. Nogami and K. Nagasaka, "Hydrolysis-Condensation of Zr-O-Si Alkoxides," *J. Ceram. Soci. of Japan*, **96**[9] 925-929 (1988).

[3] T. Lopez, M. Asamoza, L. Razo and R. Gomez, "Study of the Formation of Siliconaluminates by the Sol-Gel Method, by Means of IR, DTA and TGA," *J. Non-Cryst. Sol.*, **108** 45-48 (1989).

[4] J.B. Blum and J.W. Ryan, "Gas Chromatography Study of the Acid Catalyzed Hydrolysis of $Si(OC_2H_5)_4$," *J. Non-Cryst. Sol.*, **81**, 221-226 (1986).

[5] D.Y. Shin and S.M. Han, "Spinnability and Rheological Properties of Sols Derived from $Si(OC_2H_5)_4$ and $Zr(O-nC_3H_7)_4$ Solutions," *J. of Sol-Gel Sci. and Tech.*, **1**(3) 167-171 (1994).

[6] C. Guizard, N. Cygankiewicz, A. Labort and L. Cot, "Sol-Gel Transition in Zirconia Systems Using Physical and Chemical Processes," *J. Non-Cryst. Sol.*, **82** 86-91 (1986).

[7] M. Nogami, "Glass Preparation of the ZrO_2-SiO_2 System by the Sol-Gel Process from Metal Alkoxides," *J. Non-Cryst. Sol.*, **69** 415-423 (1985).

[8] J.P. Pirard, P. Petit, A. Mohsine, B. Michaux and F. Noville, "Silica-Zirconia Monoliths from Gels," *J. of Sol-Gel Sci. and Tech.*, **2** 875-880 (1994).

Patterned Microstructure of Sol-Gel Derived Complex Oxides Using Soft Lithography

Seana Seraji, Yun Wu, Nels Jewell-Larson, Mike Forbess, Steven Limmer, and Guozhong Cao
Department of Materials Science & Engineering
University of Washington, Seattle, WA 98195

ABSTRACT

Soft lithography has been combined with sol-gel processing to create patterned microstructures of complex oxides on silicon substrates. Piezoelectrics $Pb(ZrTi)O_3$ and $Sr_2Nb_2O_7$ were studied as model systems. It was found that dense complex oxides with the desired perovskite structures were formed after annealing at 700-800°C. Significant shrinkage accompanied the densification process. The results indicate that good wetting between the mold and the sol is critical to ensure complete filling of the capillaries and to obtain the desired structures. In addition, it was found that the conventional sol-gel process needs to be modified. The concentration of the constituents that form the solid phase after gelation may need to be increased in order to improve the final structure.

INTRODUCTION

While the semiconductor industry continues to push the limits of small-scale devices, the cost of the equipment needed to achieve such small geometries has also kept similar pace [1]. The average research lab does not have the funds to acquire, nor maintain the photolithographic and etching equipment needed to create such patterned structures. This unavailability of resources effectively prevents many groups from being able to perform research in cutting edge fields, such as MEMS technology, which require the ability to pattern a wide range of materials on a very small scale. Recently however, much attention has been given to a collection of non-photolithographic patterning techniques collectively known as Soft Lithography, which have the potential of becoming versatile and low cost methods for creating micron and sub-micron sized structures [1-9]. To date, several devices have been fabricated using soft lithographic techniques, such as polymeric FETs [2], electro-optic devices [3,4], Schottky diodes [5], silicon MOSFETs [6], and optical disk substrates [9], to name a few.

In addition to cost benefits, soft lithography has many advantages over traditional photolithography. Photolithography is very sensitive to surface topography. Since an elastomeric mold is used in soft lithography, good conformity over curved surfaces is possible, and thus non-planar substrates can be patterned with ease [1]. Another advantage is that any material that can be derived from liquid precursor can be patterned, provided that the solvent used does not swell the elastomeric mold [2]. This method is an inherently mild process [8]. As a result, many chemically and physically sensitive materials such as dyes and biomolecules can be patterned using this technique, showing again the versatility of this process [8].

As a result of the vast potential of these techniques, many reports describing the use of soft lithography to create patterned structures can be found in the open literature [1-9]. However, most of these experiments were focused on relatively simple, single component oxide systems or polymeric materials [1-9]. No work on the direct patterning of complex oxide materials has been reported in the literature. However, these materials possess many important physical properties such as ferroelectricity, piezoelectricty, pyroelectricity and high T_c superconductivity, which make complex oxides very useful for industrial and modern technological applications. For example, piezoelectric materials play a critical role in Micro-Electro-Mechanical Systems(MEMS) [10]. As such, it would be very beneficial to develop a convenient and low cost method of patterning these materials. This report describes our preliminary work on patterning complex oxide ceramics using soft lithography in conjunction with sol-gel processing. $Pb(ZrTi)O_3$ (PZT) and Strontium Niobate ($Sr_2Nb_2O_7$), both of which are piezoelectric ceramics, were chosen as model systems to form patterned structures on silicon substrates using soft lithography. Specifically, the microstructures were patterned by MIcro-Molding In Capillaries, or MIMIC molding [1-2].

PROCEDURE
Sol Preparations

In the preparation of the $Sr_2Nb_2O_7$ sol, the inorganic precursors used were strontium nitrate $Sr(NO_3)_2$ (99%) and niobium penta-chloride ($NbCl_5$) (99.8%). The procedure implemented is outlined in the flowchart, Figure 1, which differs from other reported methods [11,12]. Using ethylene glycol as a cross-linking agent and ethanol as a solvent, a transparent, stable sol was obtained. Although only 0.018 moles of water were added, hydrated citric acid was used, which provided the extra water needed to achieve the desired molar ratio of $Sr:Nb:H_2O = 1:1:3.5$. The final concentration of $Sr_2Nb_2O_7$ in the sol is of 0.6 M. This sol was used for creating the patterned structures with the soft lithography molds.

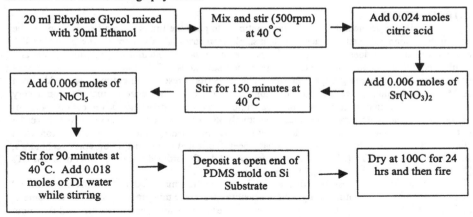

Figure 1.   Flowchart for $Sr_2Nb_2O_7$ patterning

In contrast, the PZT sol was prepared using the following organic precursors: $Pb(CH_3COO)_2$(tri-hydrate), $n-Zr(C_3H_7O)_4$, and $Ti((CH_3)_2CHO)_4$. The lead precursor was combined with acetic acid, while separately the zirconium and titanium precursors were mixed together. These two mixtures were then combined. Ethylene glycol and water were then added to this mixture. During the sol preparation, the alkoxide precursors were hydrolyzed (reacted) with H_2O to form hydroxides. Through condensation reactions, nano-sized clusters were formed [13].

Figure 2 shows the X-ray diffraction spectra of the PZT and $Sr_2Nb_2O_7$ xerogel samples derived from their respective sols. Xerogels were made from the sols, and then fired at 700-1000°C for 30 min in air. The resulting powders were then ground and used for X-ray diffraction analysis. A single phase PZT was formed, with no detectable second phases. In the $Sr_2Nb_2O_7$ system, $Sr_2Nb_2O_7$ was the major phase however a minor unidentified secondary phase was detected.

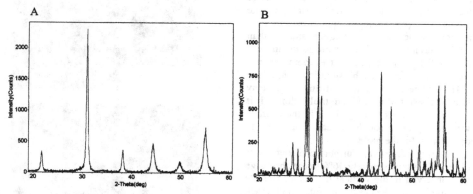

Figure 2.   XRD Spectra for (A)-PZT, (B)-$Sr_2Nb_2O_7$. Both samples from sol-gel derived powders. (* Represents unidentified phases)

Soft lithography

The process of soft lithography has been thoroughly described in many other publications, therefore only a brief description will be given here [1-2]. A positive master is first created on a silicon wafer using standard semiconductor processing techniques. The master is then covered with liquid poly-dimethylsiloxane (PDMS). When the PDMS has cured, the silicon and PDMS are separated, and thus a negative of the pattern is transferred to the PDMS. For this study, a two by two micron channel PDMS mold was made using this procedure.

The molds were then thoroughly cleaned with ethanol. The molds were then placed in conformal contact with a clean silicon substrate. The channels of the mold thus formed capillaries with the silicon substrate. The sol was then deposited at the open end of these capillaries with a transfer pipette. Through capillary force, the sol was drawn into the mold [1].

The molds and substrates were then placed in a drying oven for 24 hrs at 100°C, to drive off the solvent. As the solvent evaporated, the condensation reaction proceeded, thus promoting cross-linking between individual clusters, leading to the sol-gel transition. After the materials had gone through their sol-gel transitions, the molds were carefully removed, leaving behind positive replicas of the molds on the surface of the silicon. (PDMS has a relatively low surface energy, and therefore the material preferentially remains on the silicon substrate [2].) The patterned structures were then fired at 700 and 800°C for thirty minutes for the PZT and $Sr_2Nb_2O_7$ patterned structures respectively in order to densify the structures and to form the desired phases. The topography and morphology of these structures were then analyzed by both atomic force microscopy (AFM) and scanning electron microscopy (SEM).

RESULTS AND DISCUSSIONS

Figures 3, parts A and B, are AFM micrographs of the 2x2-micron PZT patterned structures, before and after firing, respectively. Since the concentration of the sol used was rather low, a significant amount of shrinkage was expected. The pre-firing height of the arrays was (on average) approximately 0.35 microns, which was significantly smaller than the 2 micron mold. This was not surprising, for it is well known that solvent evaporation in a wet get is accompanied by significant shrinkage [14]. The relatively low pre-firing height may also be a result of insufficient mold filling. It is possible that the capillary force in the channel was not great enough to cause the sol to completely fill the mold.

During sintering, further shrinkage was observed. Approximately 46% linear shrinkage was observed in the vertical direction. Despite the significant shrinkage, uniform and crack free patterned structures were obtained, as can be seen from SEM picture (Figure 3-C). Continuous grooves running along the tops of the channels were also observed after firing. The cause of this is debatable; however it is believed that the groove formation is a result of a density gradient, and the slip-casting effect. When the sol fills the molds, the solvent will in part permeate through the mold to evaporate, causing a build up of solid material along the walls of the mold [15]. This can result in a structure that has a shell more dense than the inner material. As the material was fired, the top of the structure collapsed, due to the greater extent of shrinkage that occurred in the more porous inner gel, thus causing the observed groove to develop along the top.

Figure 4, A and B are AFM micrographs of unfired and fired 2x2-mircon $Sr_2Nb_2O_7$ channels, respectively. The initial height of this pattern (0.673 microns) was also found to be significantly smaller than the 2-micron mold. The $Sr_2Nb_2O_7$ sol was much more viscous than

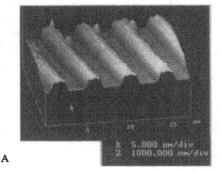

A

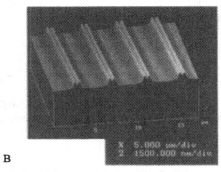

B

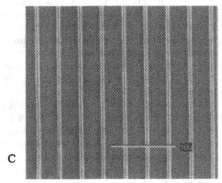

C

Fig. 3 A-PZT sample before firing (AFM)
B-PZT sample after firing (AFM)
C-PZT sample after firing (SEM)

the PZT, and thus might not have dried completely after 24 hours at 100°C. As a result, the unfired samples appear to be somewhat larger than the PZT samples. When fired though, this sample exhibited shrinkage of approximately 85% in the vertical direction whereas the horizontal shrinkage of this sample was negligible. This shrinkage behavior is significantly different from that of the PZT sample. This phenomenon might imply that no dense shell was formed during drying. Figure 4-C is a SEM micrograph of the strontium niobate sample and indicates that no cracks were formed.

Another observation made involved the fidelity of the mold replication. The molds used for this study were all rectangular with sharp 90° side-walls. However, as can be seen from Figures 3 and 4, the side-walls of the resulting patterns were tilted to varying degrees. This effect was even more pronounced after firing. One possible explanation is that the bottom edge of the channel was constrained by its adhesion to the silicon substrate, while the top of the channel was free of constraints. Thus, the top was free to shrink in any direction, while the bottom could not contract in the horizontal direction, which would give rise to a slanting of the walls of the channels.

In summary, a procedure for patterning complex, multi-component piezoelectric ceramics (with PZT and $Sr_2Nb_2O_7$ as a model system) using micro-capillary molding in conjunction with sol-gel techniques was described. Crack free and uniform microstructures were obtained and the desired crystalline phases were formed after firing at high temperatures. However, significant shrinkage was found in the patterned structures prior to and after firing. Such shrinkage resulted in not only a dimension change, but also a distortion of the shape of the structure. A higher load of solid constituents should reduce the shrinkage. In addition the formation of a dense outer shell seems to be effective in minimizing shape distortion.

ACKNOWLEDGMENT:

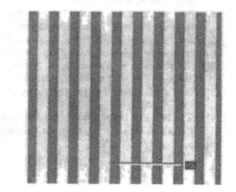

Fig. 4 A- $Sr_2Nb_2O_7$ sample before firing (AFM)
B- $Sr_2Nb_2O_7$ sample after firing (AFM)
C- $Sr_2Nb_2O_7$ sample after firing (SEM)

The authors of this paper would like to thank Dong Qin at the Washington Technologies Center for her indispensable assistance with the atomic force microscope.

REFERENCES

[1] Y. Xia, G.M. Whiteside, Soft Lithography, *Annu. Rev. Mater. Sci* **28**, 1998, 153.
[2] W.S. Beh, I.T. Kim, D. Qin, Y. Xia, G.M Whitesides, Formation of patterned microstructures of conducting polymers by soft lithography and applications in microelectronic device fabrication. *Adv. Mater.* **11**, [12], 1999, 1038
[3] J.T. Rantala, G.E. Jabbour, J. Vahakanagas, S. Honkanen, B. Kippelen, N. Peyghambarian, Hybrid sol-gel micro-patterning of organic electroluminescent devices, *Jap. J. Applied Phys.*, Part 2(Letters), 37 [no 10A, 1], 1998, L1098.
[4] J.A. Rogers, M. Meier, A. Dodabalaour, Distributed feedback lasers produced using soft lithography, *56th An. Device Res. Conf. Dig. IEEE*, New York, NY, USA; 1998, 134.
[5] J. Hu, R.G. Beck, R.M. Westervelt, G.M. Whitesides, Use of soft lithography to fabricate arrays of Schottky diodes, *Adv. Mater.* **10**, [8], June 1998, 574
[6] N.L. Jeon, J. Hu, G.M. Whitesides, Fabrication of silicon MOSFETs using soft lithography, M. K. Erhardt, R.G. Nuzzo, *Adv. Mater.* **10**, [17], Dec 1998, 1466.
[7] M.J. Lochhead, P. Yager, Multi-component micro-patterned sol-gel materials by capillary molding, *Intl. Soc. for Optical Engr.*, 3136, 1997, 261.
[8] E. Kim, Y. Xia, G.M. Whiteside, Two- and three-dimensional crystallization of polymeric microspheres by micromolding in capillaries, *Adv. Mater.* Vol.8, No.3; March 1996; 245.
[9] A. Matsuda, Y. Matsuno, Y. Mitsuhashi, N. Tohge, T. Minami, Optical disk substrate fabricated by the sol-gel method, Key Engineering Materials v 150, 1998, 111-120
[10] D.L. Polla, L.F. Francis, Ferroelectric thin films in micro-electromechanical systems applications, *MRS Bulletin*, July 1996, 59
[11] A.V. Prasadaro, U. Selvaraj, S. Komarneni, Fabrication of $Sr_2Nb_2O_7$ thin films by sol-gel processing, *J. Mater. Res.*, **10**, [3], March 1995, 704.
[12] K. Okuwada, S. Nakamura, H. Nozawa, Crystal growth of layered perovskite $Sr_2Nb_2O_7$ $Sr_2Ta_2O_7$ film by the sol-gel technique, *J. Mater. Res.*, **14**, [3], March 1999, 855.
[13] G. Yi, M. Sayer, Sol-gel processing of complex oxide films, *Ceramic Bulletin*, **70**, [7], 1991, 1173.
[14] C. J. Brinker, G.W, Scherer, *Sol-Gel Sciences*, Academic Press Inc, San Diego, Ca, 1990.
[15] J.S. Reed, *Principles of Ceramic Processing*, Wiley, John & Sons, Inc., 1994

$SrBi_2(Nb,V)_2O_9$ FERROELECTRIC FILMS BY SOL-GEL PROCESSING

Y. Wu, S. Seraji, M.J. Forbess, S.J. Limmer, and G.Z. Cao
University of Washington
Materials Science and Engineering
302 Roberts Hall, Box 352120
Seattle, WA 98195

ABSTRACT

Our recent experiments demonstrated that $SrBi_2(Nb, V)_2O_9$ ceramics possess excellent ferroelectric properties. Specifically, partial substitution of niobium by vanadium resulted in an appreciably reduced coercive force, from 63 kV/cm without doping to 45 kV/cm with 10 at% vanadium doping, and a significant increase in remanent polarization, from 2.8 $\mu C/cm^2$ without doping to 8.0 $\mu C/cm^2$ with doping. $SrBi_2(Nb,V)_2O_9$ ferroelectrics are a promising candidate for applications in non-volatile random access memories (NvRAMs); however, the ability of making $SrBi_2(Nb, V)_2O_9$ films is essential for such application. In this paper, we report synthesis and ferroelectric properties of $SrBi_2(Nb, V)_2O_9$ films by sol-gel processing. A sol-gel processing route has been developed using inorganic salts as precursors, except for vanadium. XRD indicated that single phase layered perovskite ferroelectric sol-gel films with vanadium doping up to 10 at% were obtained after a heat-treatment at 600-800 °C. Preliminary experiments indicate that the vanadium doping exhibits similar influences on the ferrolectric properties of the sol-gel films.

INTRODUCTION

Ferroelectric thin films are promising candidates for non-volatile memories in digital information storage systems as ferroelectric random access memories (FeRAMs) [1-2]. Recently bismuth layered perovskite $SrBi_2(Ta,Nb)_2O_9$ (SBTN) materials based on $SrBi_2Ta_2O_9$ (SBT) and $SrBi_2Nb_2O_9$ (SBN) have attracted increasing attention due to their excellent fatigue resistance and ferroelectric properties comparable to those of $Pb(Zr,Ti)O_3$ (PZT) [3]. There was no significant change of remanent polarization after switching of 2×10^{11} cycles in SBTN system [4]. However, layered perovskite ferroelectric films suffer from the drawback of a relatively low remanent polarization [5]. Efforts have been made to enhance the properties of layered perovskite ferroelectrics by the addition or substitution of alternative cations [6]. For example, partial substitution of Sr^{2+} by Bi^{3+} has resulted in the most noticeable improvement of ferroelectric properties [7-9].

In our previous work [10-12], we have reported the significant enhancement of ferroelectric properties of SBN ferroelectric ceramics through partial substitution of niobium by pentavalent vanadium cations. With partial substitution of niobium by vanadium cations (up to 30 at%), the single-phase layered perovskite structure was preserved and the sintering temperature of the ceramic system was significantly lowered (~ 200 °C). The incorporation of vanadium into the layered perovskite structure resulted in a shift of the Curie point to higher temperatures from 418 °C to ~430 °C with 10 at% vanadium doping, and an increase in dielectric constant, from ~ 700 to ~ 1200 with 10 at% vanadium doping, at their respective Curie points. The remanent polarization

---

To the extent authorized under the laws of the United States of America, all copyright interests in this publication are the property of The American Ceramic Society. Any duplication, reproduction, or republication of this publication or any part thereof, without the express written consent of The American Ceramic Society or fee paid to the Copyright Clearance Center, is prohibited.

increased from ~ 2.4 µC/cm^2 to ~ 8 µC/cm^2, while the coercive field decreased from ~ 63 kV/cm to ~ 45 kV/cm with 10 at% V^{5+} doping. The above results clearly indicate a significant increase in polarizability with an increased doping of vanadium. This is attributed to the increased "rattling space" due to the incorporation of much smaller vanadium cations.

For many applications, including FeRAMs, the ability of making SBVN thin films is essential. It has been reported that the bismuth layered perovskite films were made by sputter deposition [13], pulsed laser-ablation [1] and sol gel processing [14]. Sol-gel method has the advantage of low cost and easy handling of the starting materials. Sol-gel process is also compatible to patterning with device dimension not less than 2 µm. In this paper, we report on the synthesis of SBVN layered perovskite films deposited on (100) single crystal silicon substrates by sol-gel processing and the influence of vanadium doping on the ferroelectric properties of SBVN films.

RESULTS AND DISCUSSION

The SBVN sol was prepared by using inorganic salts as the precursors except vanadium. Strontium nitrate Sr(NO$_3$)$_2$ (99+%, Aldrich) and bismuth nitrate pentahydrate, Bi(NO$_3$)$_3$·5H$_2$O, (98+%, Aldrich) were used for the sol preparation without any pre-treatment, while the niobium precursor was prepared by dissolving niobium chloride, NbCl$_5$, (99%, Aldrich) in excess dehydrated alcohol in a dry nitrogen environment. The niobium chloroalkoxide was formed by the following reaction:

$$NbCl_5 + 3ROH \text{ (excess)} \rightarrow NbCl_2(OR)_3 + 3HCl \quad (1)$$

Then the solution (0.3 M) was kept stirring for two hours under nitrogen condition to remove the hydrogen chloride gas. The vanadium (V) tri-isopropoxide oxide (95-99%, Chemat) was used as the precursor for vanadium. It was diluted by isopropanol to 0.2 M under dry nitrogen condition. The strontium and bismuth stock solutions were prepared by separately dissolving Sr(NO$_3$)$_2$ and Bi(NO$_3$)$_3$·5H$_2$O into a mixture of ethylene glycol (EG) with citric acid and alcohol under stirring (600 RPM) at room temperature for two hours. Mixtures of EG and alcohol with different volume ratios were tried and it was found that both Sr(NO$_3$)$_2$ and Bi(NO$_3$)$_3$·5H$_2$O could be best dissolved into a mixture with a volume ratio of 2:3 of EG and alcohol. The molar ratios of Sr(NO$_3$)$_2$:CA and Bi(NO$_3$)$_3$:CA in stock solutions are 1:2 and 2:1, respectively. The molar concentrations of the strontium and bismuth stock solutions were adjusted to be 0.225 M and 0.3 M and an 8 wt% excess Bi content was added into each coating sol to compensate for the bismuth loss during the heat-treatment. During the final part of preparation for coating solution, the specific amounts of strontium and bismuth stock solutions were mixed together first. After adding and dissolving the remaining amount of CA to the mixed solution to reach the 2:1 molar ratio of citric acid to all the metal ions, the niobium and vanadium stocking solutions were added into it to get the coating solution. After stirring 2 more hours, stable transparent sols were obtained and ready for coating. The pH values of the sols were approximately 2 and the resultant sol is stable for more than one month when stored at room temperature.

SBVN thin films on single crystal (100) Si wafers were prepared by spin-coating of the sol. The substrates were cleaned in ultrasonic bath with acetone and alcohol. The SBVN sols were spin-deposited onto silicon substrates at 4000 rpm for 40 seconds. Between each coating, the samples were put onto a hot plate (around 300 °C) for 10 minutes to get rid of the residual organic solvent. Typically, 5-8 coatings were applied to get SBVN films for characterization and property measurement. After the final coating, the substrates were heat-treated at 800 °C for 1 hr in ambient air condition with a heating rate of about 80 °C/min. The electrical measurements were conducted on films in a metal-ferroelectric-insulator-semiconductor (MFIS) configuration using Ag as the gate electrode and the back contact.

Fig.1. shows the XRD results of SBVN films on (100) silicon wafers by using Philips PW 1830 XRD generator with Cu Kα radiation. The XRD results show after a heat-treatment at 800 °C, a single phase layered perovskite structure was formed and no secondary phase was detected regardless of vanadium doping levels up to 20 at%. Although the size of pentavalent vanadium ions is significantly smaller than that of Nb^{5+}, the lattice constants underwent a negligible reduction, similar to that found in the bulk ceramics. This negligible change in the crystal lattice was attributed to the crystal structure constraint induced by the $Bi_2O_2^{2-}$ inter-layers.

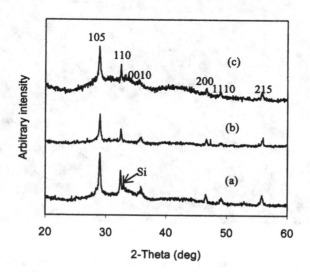

Figure 1. XRD spectra of (A) SBN, (B)SBVN (5 at% V), and (C) SBVN (10 at% V) films on silicon (100).

Fig.2. compares the XRD results of SBN powder and SBN films deposited on (100) silicon wafer and (100) $SrTiO_3$ substrates and fired at 800 °C under ambient atmosphere for 1 hr. The film on (100) $SrTiO_3$ shows very strong c-orientation because of the very small lattice mismatch. Desu etc reported the ferroelectric properties of c-oriented SBN films [15]. The films are quite uniform and crack-free (see Fig.3). All of the films show dense structure with an average grain size of approximately 100 nm.

Fig.3. shows the capacitance verse temperature plots of SBVN films. From the results, the films show similar changing trend of capacitance over temperature. Also the Curie temperatures are very close to the respective bulk ceramics. More detailed systematic experiments are required to achieve a better understanding of the relationship between the vanadium doping and ferroelectric properties of layered perovskite ferroelectric films.

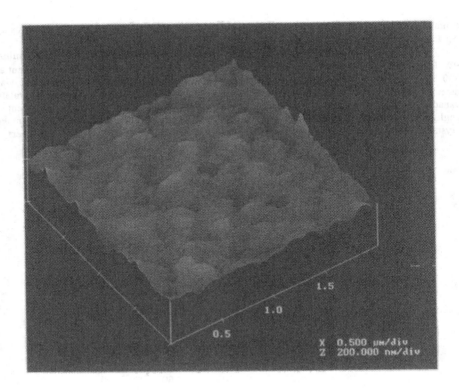

Figure 2.   3-D mode AFM picture of SBVN film on silicon wafer.

CONCLUSIONS

In summary, layered perovskite strontium bismuth niobate ferroelectric films doped with vanadium were fabricated by sol gel processing. Dense and smooth thin films were obtained on silicon wafer. The layered structures ferroelectric properties of SBN thin films were kept at least up to 10 at% vanadium doping. The influence of vanadium doping might be due to the larger "rattling space" of vanadium ions inside the octahedra. Further analysis of dielectric and ferroelectric properties of SBVN films is in progress.

ACKNOWLEDGMENT:

The authors would like to acknowledge valuable technical support of Dave Rice and Yadong Yin. One of the authors (Y. Wu) would like to thank the Center of Nanotechnology at University of Washington for the financial support as a NanoTech fellow.

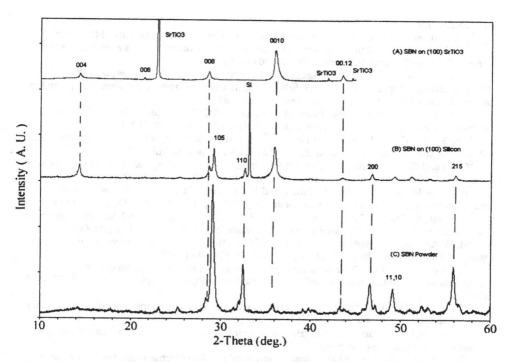

Figure 3. XRD spectra of SBN powder and SBN films on silicon (100) and SrTiO$_3$ (100).

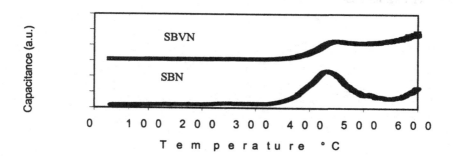

Figure 4. Capacitance values of the film capacitors vs temperature. SBN (●), and SBVN(○, with 5 at% V) as a function of temperature, measured at a frequency of 100 kHz.

REFERENCES

[1] J.F. Scot and C.A.P. de Araujo, "Ferroelectric memories," Science **246**, 1400-1405 (1989).

[2] J.F. Scott, "High-dielectric constant thin films for dynamic random access memor(DRAM)," Annual Review of Materials Science, **28**, 79-100 (1998).

[3] P. Y. Chu, R. E., Jones Jr., P. Zurcher, D. J. Taylor, B. Jiang, S. J. Gillespie, and Y. T. Lii, "Characteristics of spin-on ferroelectric $SrBi_2Ta_2O_9$ thin film capacitor s for ferroelectric random access memory applications", J. Mater. Res., **11**, 1065-68 (1996).

[4] C. A-Paz de Araujo, L.D. McMillan, J.D. Cuchiaro, M.C. Scott, and J.F. Scott , "Fatigue-free ferroelectric capacitors with platinum electrodes," Nature **374**, 627-629 (1995).

[5] J.F. Scott, "Thin Film Ferroelectric Materials and Devices", pp.115 in Layered Perovskite Thin Films and Memory Devices, ed. R. Ramesh, Kluwer, Norwell, MA, 1997.

[6] C. Lu and C. Wen, Mater. "Preparation Properties of Barium Incorporated Strontium Bismuth Tantalate Ferroelectric Thin Film," Res. Soc. Symp. Proc. **541**, 229-234 (1999).

[7] P. Duran-Martin, A. Castro, P. Millan, and B. Jimenez, "Influence of Bi-site substitution on the ferroelectricity of the Aurivillius compound $Bi_2SrNb_2O_9$," J. Mater. Res. 13, 2565-71 (1998).

[8] Y. Torii, K. Tato, A. Tsuzuki, H. J. Hwang, S. K. Dey, "Preparation and dielectric properties of nonstoiometric $srBi_2Ta_2O_9$-based ceramics", J. Mater. Sci. Lett., **17**, 827-829 (1998).

[9] H. Watanabe and T. Mihara, "Preparation of ferroelectric thin films of bismuth layer structured compounds," Jpn. J. Appl. Phys., Part I, **34**, 5240-44 (1995).

[10] Y. Wu and G.Z. Cao, "Enhanced ferroelectric properties and lowered processing temperatures of strontium bismuth noibates with vanadium doping," Appl. Phys. Lett. **75**, 2650-52 (1999).

[11] Y. Wu and G.Z. Cao, "Influences of vanadium doping on ferroelectric properties of strontium bismuth niobates," J. Mater. Sci. Lett., **15**, 267-269 (2000).

[12] Y. Wu and G. Z. Cao, "Ferroelectric and dielectric properties of strontium bismuth niobate vanadates", J. Mater. Res. (in press).

[13] T. K. Song, J.-K. Lee, and H. J. Jung, " Structural and ferroelectric properties of the c-axis oriented $SrBi_2Ta_2O_9$ thin films deposited by the radio-frequency magnetron sputtering", Appl. Phys. Lett. **69**, 3839-41 (1996).

[14] K. Kato, C. Zheng, J.M. Finder, and S.K. Dey, "Sol-gel route to ferroelectric layer-structured perovskite $SrBi_2Ta_2O_9$ and $SrBi_2Nb_2O_9$ thin films," J. Am. Ceram. Soc. **81**, 1869-1875 (1998).

[15] S.B. Desu and D.P. Vijay, "C-axis oriented ferroelectric $SrBi_2(Ta_x, Nb_{2-x})O_9$ thin films," Mater. Sci. Eng. **B32**, 83-88 (1995).

FLEXIBLE SHEETS OF DIMETHYLSILOXANE-BASED INORGANIC/ ORGANIC HYBRIDS

Shingo Katayama, Keiko Kawakami and Noriko Yamada
Nippon Steel Corporation, Advanced Technology Research Laboratories,
20-1 Shintomi, Futtsu, Chiba 293-8511, JAPAN

ABSTRACT

Large sheets of dimethylsiloxane-based inorganic/organic hybrids with a size of 200 x 290 mm and about 1.5 mm thickness have been successfully fabricated from polydimethylsiloxane (PDMS) and chemically modified metal alkoxides. The hybrid sheets were flexible, homogeneous and transparent without inorganic precipitates from metal alkoxides. They showed high elongation of 100 % in the hybrid sheets fabricated in $Zr(OBu^n)_4$/PDMS = 2 at 180 °C and high strength of 3.0 MPa in those fabricated in $Zr(OBu^n)_4$/PDMS = 4 at 180 °C. The storage modulus of 10^{7-8} Pa in rubbery region was approximately constant at temperatures of 0-300 °C, possibly having high temperature stability in flexibility.

INTRODUCTION

Inorganic/organic hybrid materials, in which inorganics and organics are combined at a molecular level by the sol-gel process, are attracting attention because of the possibility for providing both properties of inorganic and organic materials. Dimethylsiloxane-based hybrids prepared from tetraethoxysilane (TEOS) and polydimethylsiloxane have been intensively studied, resulting in it being determined that they have unique mechanical properties[1-4]. It has become clear that the mechanical properties and structure of the hybrids are strongly influenced by the reaction conditions such as acid concentration, TEOS content and molecular weight of PDMS[2,4]. Mackenzie et al. reported the rubbery behaviors of dimethylsiloxane-based hybrids "Ormosils", based on the viewpoint in which inorganic oxides were modified by the incorporation of organic groups to improve the ductility[3,4]. It is possible that they may be thermally more stable than pure organic rubbers, because the inorganic component derived from metal alkoxides is incorporated and may act as a thermally stable cross-linking agent of PDMS. However, the Ormosils have the TEOS-derived siloxane network as an inorganic component. The use of various metal alkoxides instead of TEOS is anticipated to alter the structure and properties of dimethylsiloxane-based hybrids, reflecting the difference in molecular weight, valence and coordination number among silicon and other metallic elements.

The authors have examined metal alkoxides other than Si alkoxides such as TEOS in dimethylsiloxane-based hybrids. The inorganic components have homogeneously been incorporated into the dimethylsiloxane-based hybrids by chemical modification of metal alkoxides with ethyl acetoacetate[5-8]. The fabrication of large sheets of the dimethylsiloxane-based hybrids has been further investigated toward their practical applications as thermally stable rubbers.

In this paper, the fabrication and mechanical properties of flexible sheets of dimethylsiloxane-based hybrids from PDMS and chemically modified metal alkoxides are described.

EXPERIMENTAL

Zirconium n-butoxide $Zr(OBu^n)_4$ and tantalum ethoxide $Ta(OEt)_5$ were used as precursors of inorganic components, which were commercially available from KANTO CHEMICAL CO., INC. and NIHON KAGAKU SANGYO CO., LTD., respectively. Liquid PDMS with an average molecular weight of 3000 was purchased from Shin-Etsu Chemical Co., Ltd. 2-ethoxyethanol used as solvent. Ethyl acetoacetate (EAcAc) was used as a chemical modifier of the metal alkoxides in order to prevent the formation of precipitates in hydrolysis of the metal alkoxides. Metal alkoxide and EAcAc were mixed in a metal alkoxide : EAcAc molar ratio of 1 : 2, followed by addition of 2-ethxyethanol and PDMS. A mixture of water and 2-ethoxyethanol was added to the 2-ethoxyethanol solution of metal alkoxide and PDMS. The molar ratio of metal alkoxide, PDMS, EAcAc, H_2O and ethanol was 1 : 0.25-0.5 : 2 : 2 : 4. After stirring the mixture for 30 min, the solution was poured into a polytetrafluoroethylene-coated SUS vat with 135 x 175 mm, 190 x 225 mm and 210 x 304 mm sizes and covered with aluminum foil. The solution was allowed to gel at 100 °C for 1 day then heat-treated at 160-180 °C for 3 days to produce a dimethylsiloxane-based hybrid.

Stress-strain experiments were carried out at room temperature with an Instron load cell. Dumbbell-shaped samples (dumbbell-type No.3, JIS K 6301) were used. The crosshead speed was 1 mm/min.

The dynamic mechanical data were obtained with a dynamic viscoelastometer of Seiko Instruments Inc. The specimen size was about 30 x 7 x 1.5-2.0 mm. Samples were tested at 10 Hz in the temperature range of –150 to 300 °C with a heating rate of 3 °C/min.

RESULTS AND DISCUSSION

Figure 1 shows a typical sample of dimethylsiloxane-based hybrid sheets fabricated from PDMS and chemically modified $Zr(OBu^n)_4$, which is the largest size of 200 x 290 mm with about 1.5 mm thickness among our hybrid sheets. The hybrid sheets were flexible, homogeneous and transparent without inorganic particles precipitated by hydrolysis of metal alkoxides. Even by TEM, no particular long-range order such as oxide particles with 10 nm size or above was observed in the hybrid sheets. It is likely that the inorganic component derived from metal alkoxides in the hybrid sheets is close to the molecular-level size and homogeneously distributed at its size, as reported previously by the authors[7].

The mechanical properties of dimethylsiloxane-based hybrid sheets fabricated at heat-treatment temperatures of 160-180 °C are shown in Fig. 2-4. The tensile strength increased with increasing the heat-treatment temperature as shown in Fig. 2. The hybrid

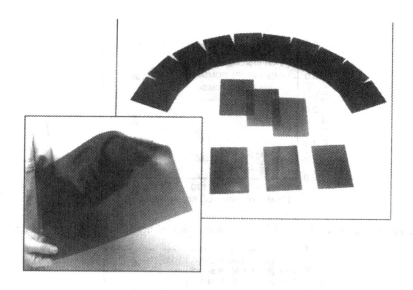

Fig. 1 Dimethylsiloxane-based inorganic/organic hybrid sheets.

sheets fabricated at 180 °C exhibited the highest strength of 1.5-3.0 MPa, although it was reported that the strengths of flexible TEOS-PDMS hybrids were around 0.2MPa[9] and 1.2 MPa[2]. The Young's modulus of the dimethylsiloxane-based hybrid sheets also increased with increasing the heat-treatment temperature as shown in Fig. 3. The increase in Young's modulus in $Zr(OBu^n)_4$/PDMS = 4 was lager than that in $Zr(OBu^n)_4$/PDMS = 2. Figure 3 also indicates that higher content of inorganic components provided higher Young's modulus, as reported previously[7]. The elongation at break was insignificantly varied with the heat-treatment temperature at a high inorganic content of $M(OR)_n$/PDMS = 4, showing around 5-20 %. These values are similar to those of hybrids with the inorganic component of silicate derived from TEOS. However, the hybrid sheets fabricated at $Zr(OBu^n)_4$/PDMS = 2 exhibited the dependence of elongation at break on heat-treatment temperature. The largest elongations above 100 % were at a heat-treatment temperature of 180 °C. Although the hybrid sheets had high Young's modulus at the high heat-treatment temperatures and were hard to be distorted, they provided such a large elongation. The large elongation at high temperatures of the heat-treatment is likely due to the increase in tensile strength. Thus, it is difficult for the hybrids to break even at larger elongation.

Metal alkoxides such as $Zr(OBu^n)_4$ and $Ta(OEt)_5$ other than TEOS can provide large elongation in dimethylsiloxane-based hybrids. They are thought to act as a strong cross-linking agent of PDMS chains, compared with TEOS. In addition, because it was described that self-condensation of some PDMS chains takes place in the TEOS-PDMS system, the metal alkoxides may act as an effective catalyst of the self-condensation reaction and make stronger hybrid sheets.

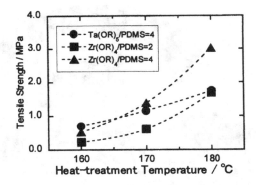

Fig. 2 Tensile strength of hybrid sheets as a function of heat-treatment temperature.

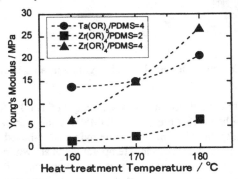

Fig. 3 Young's modulus of hybrid sheets as a function of heat-treatment temperature.

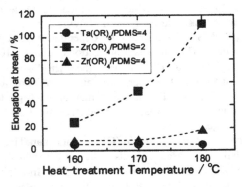

Fig. 4 Elongation at break of hybrid sheets as a function of heat-treatment temperature.

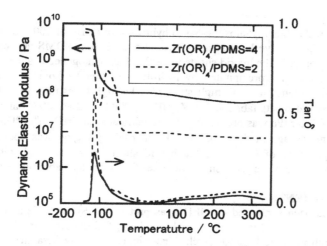

Fig. 5 Storage modulus and tan δ of hybrid sheets fabricated in $Zr(OBu^n)_4$/PDMS = 2 and 4 at 180 °C.

The flexibility of the hybrid sheets was also evaluated by dynamic mechanical measurements. The storage modulus and tan δ of the hybrid sheets fabricated in $Zr(OBu^n)_4$/PDMS = 2 and 4 at 180 °C are shown in Fig. 5. The storage modulus at temperatures of 0-300 °C was approximately constant in rubbery region. The hybrid sheets may have high temperature stability in flexibility. The abrupt decrease in storage modulus and the peak of tan δ around -100 °C are attributed to the glass transition of PDMS chains. Regarding the glass transition of PDMS chains in the hybrid sheets, the high inorganic content of $M(OR)_n$/PDMS = 4 showed a smaller decrease in storage modulus and a broader peak of tan δ than the low inorganic content of $M(OR)_n$/PDMS = 2. Furthermore, in the hybrid fabricated in $Zr(OBu^n)_4$/PDMS = 2, the peak of storage modulus around -80 °C results from crystallization and melting of PDMS chains, which were also confirmed by DSC analysis. These results reveal that the hybrid sheets with a low inorganic content involves a number of free PDMS chains and/or PDMS chains with many degrees of freedom, giving very high elongation.

The inorganic components derived from Zr and Ta alkoxides are close to the molecular-level size and behave as strong cross-linking agents of PDMS chains. It is different from the dispersion of PDMS chains within the silicate network derived from TEOS[2]. Therefore, the hybrid sheets fabricated from PDMS and metal alkoxides of Zr and Ta are thought to have a flexible structure in which flexible PDMS chains are strongly cross-linked with the cross-linking agent of inorganic components derived from metal alkoxides, resulting in higher elongation and higher strength.

CONCLUSIONS

Large sheets of the dimethylsiloxane-based hybrids with a size of 200 x 290 mm and about 1.5 mm thickness have been successfully fabricated from PDMS and chemically modified metal alkoxides. The hybrid sheets were flexible, homogeneous and transparent without inorganic particles precipitated by hydrolysis of metal alkoxides. The hybrid sheets showed high elongation and high strength. The high elongation of 100 % was observed in the hybrid sheets fabricated in $Zr(OBu^n)_4$/PDMS = 2 at 180 °C and the high tensile strength of 3.0 MPa was observed in those fabricated in $Zr(OBu^n)_4$/PDMS = 4 at 180 °C. Since the storage modulus of 10^{7-8} Pa in rubbery region was approximately constant at temperatures of 0-300 °C, the hybrid sheets may have high temperature stability in flexibility. They are promising candidates for thermally stable rubbers.

ACKNOWLEDGMENTS

This work has been supported by NEDO, as part of the Synergy Ceramics Project promoted by AIST, MITI, Japan. Tow of the authors, Shingo Katayama and Noriko Yamada, are members of the Joint Research Consortium of Synergy Ceramics.

REFERENCES

[1] G. L. Wilkes, B. Orler and H. Huang, "Ceramer: Hybrid Materials Incorporating Polymeric/Oligomeric Species into Inorganic Glasses utilizing a Sol-Gel Approach," *Polym. Prep.* **26**, 300-302(1985).

[2] H. Huang, B. Orler and G. L. Wilkes, "Structure-Properties Behaivio of New Hybrid Materials Incorporating Oligomeric Species into Sol-Gel Glasses. 3. Effect of Acid Content, Tetraethoxysilane Conetnt, and Molecular Weight of Poly(dimethylsiloxane)," *Macromolecules* **20**, 1322-1330(1987).

[3] Y. J. Chung, S. –J. Ting and J. D. Mackenzie, "Rubbery Ormosils," *Mat. Res. Soc. Proc.* **180**, 981-986 (1990).

[4] J. D. Mackenzie, Q. Huang and T. Iwamoto, "Mechanical Properties of Ormosils," *J. Sol-Gel Sci. Technol.* **7**, 151-161(1996).

[5] S. Katayama, I. Yoshinaga and N. Yamada, "Synthesis of Inorganic-Organic Hybrids from Metal Alkoxides and Silanol-terminated Polydimethylsiloxane," *Mat. Res. Soc. Symp. Proc.* **435**, 321-326(1996).

[6] N. Yamada, I. Yoshinaga and S. Katayama, "Synthesis and Dynamic Mechanical Behavior of Inorganic-Organic Hybrids Containing Various Inorganic Components," *J. Mater. Chem.* **7[8]**, 1491-1495(1997).

[7] N. Yamada, I. Yoshinaga and S. Katayama, "Effect of Inorganic Components on the Mechanical Properties of Inorganic-Organic Hybrids Synthesized from Metal Alkoxides and Ploydimethylsioxane," *J. Mater. Res.* **14**, 1720-1726(1998).

[8] N. Yamada, I. Yoshinaga and S. Katayama, "Processing and Properties of Inorganic-Organic Hybrids Containing Various Inorganic Components," *J. Sol-Gel Sci. Tech.* **13**, 445-449(1998).

[9] J. D. Mackenzie, Q. Huang, F. Rubio-Alonso and S. J. Kramer, "Effects of Temperature on Properties of Ormosils," *Mat. Res. Soc. Symp. Proc.* **435**, 229-236(1996).

ACHIEVEMENT OF CRACK-FREE CERAMIC COATINGS OVER 1 μm IN THICKNESS VIA SINGLE-STEP DEPOSITION

Hiromitsu KOZUKA,* Katsumi KATAYAMA, Yoshiro ISOTA and Shinsuke TAKENAKA
Dept. Mater. Sci. Eng., Kansai Univ., 3-3-35 Yamate-cho, Suita, 564-8680, Japan

ABSTRACT

$BaTiO_3$ and PZT coating films were prepared by dip- or spin-coating from metal alkoxide solutions containing polyvinylpyrrolidone (PVP). The gel film deposition and firing were performed just one time, not repeated, and the critical thickness, the maximum thickness achievable without crack formation, was evaluated by varying the gel film thickness via aging the sols or via changing the substrate withdrawal or rotation speed. The critical thickness significantly increased when PVP was added in the starting solutions. X-ray diffraction measurements also demonstrated that PVP enhances the formation of single-phase $BaTiO_3$ and PZT on heat-treatment of the gel films. Crack-free $BaTiO_3$ and PZT films 2.1 and 1.7 μm in thickness were obtained, respectively, via single-step deposition, and the thick PZT film thus obtained showed polarization-electric field hysteresis.

INTRODUCTION

Sol-gel coating technique has been applied to fabrication of various functional ceramic coatings in laboratories. When crack-free coatings over submicrometer are needed in laboratories, the gel film deposition and firing are repeated using coating solutions of low viscosities in order to avoid crack formation. The maximum thickness achievable without cracks via non-repetitive, single-step deposition is often called *critical thickness*, which is usually less than 0.1 μm for non-silicate polycrystalline films. The low values of the critical thickness could be one of the factors that discourage people in industries from employing the sol-gel technique in mass production of polycrystalline ceramic coatings.

So far, chelating agents [1] and diols as solvents [2] have been reported to be effective in enhancing the critical thickness. Recently the present authors have shown that crack-free $BaTiO_3$ [3-6] and PZT [6] films both 1.2 μm in thickness can be prepared via single-step deposition from alkoxide-derived sols containing polyvinylpyrrolidone (PVP, Fig. 1). In $BaTiO_3$ films the residual stress was found to reduce significantly by incorporating PVP in the solutions [5,6], where PVP was thought to block the condensation sites of the metalloxane polymers, promoting the stress relaxation in films on firing.

The effect of PVP on the critical thickness of $BaTiO_3$ was quantitatively evaluated in the present paper. Achievement of crack-free $BaTiO_3$ and PZT films 2.1 and 1.7 μm in thickness via single-step deposition is also reported. The effect of PVP on the formation of $BaTiO_3$ and PZT phases was investigated by X-ray diffraction measurement.

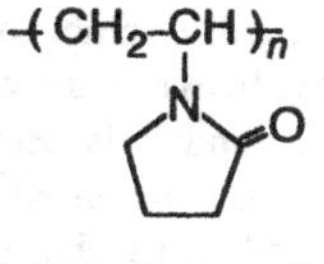

Fig. 1 Polyvinylpyrrolidone.

To the extent authorized under the laws of the United States of America, all copyright interests in this publication are the property of The American Ceramic Society. Any duplication, reproduction, or republication of this publication or any part thereof, without the express written consent of The American Ceramic Society or fee paid to the Copyright Clearance Center, is prohibited.

EXPERIMENTAL

PVP powders of average molecular weight of 630000 were employed. For fabricating BaTiO$_3$ films, starting solutions of molar compositions, Ba(CH$_3$COO)$_2$: Ti(OC$_3$H$_7^i$)$_4$: PVP : CH$_3$COOH : H$_2$O : i-C$_3$H$_7$OH = 1 : 1 : 0 - 1 : 9.08 : 20 : 20, were prepared, following the procedure described in [4]. The moles of PVP represent those of the monomer (polymerization repeating unit) of PVP. The solutions were kept standing at 30°C in a sealed glass container for various periods of time and served as coating solutions. For preparing PZT films starting solutions of molar composition, Pb(CH$_3$COO)$_2$·3H$_2$O : Zr(OC$_3$H$_7^i$)$_4$: Ti(OC$_3$H$_7^i$)$_4$: PVP : CH$_3$COOH : H$_2$O : CH$_3$OCH$_2$CH$_2$OH : n-C$_3$H$_7$OH = 1.1 : 0.53 : 0.47 : x : y : 2 : 10 : 1.9 were prepared where $x = 0.5 - 1.0$ and $y = 5$ or 10, following the procedure reported elsewhere [5]. The solution was kept standing at 25°C in a sealed glass container for 50 h and served as coating solutions. Both for preparing BaTiO$_3$ and PZT films, gel films were deposited on silica glass substrates (ca. 20 x 40 x 1.2 mm^3) either by dip- or spin-coating. Gel films were fired by transferring into an electric furnace of 700°C and heating there for 10 min. The gel film deposition was performed just one time, not repeated.

The viscosity of the sol was measured with a rotating viscometer (Brookfield, DV-I+). The film thickness was measured by a Kosaka Laboratory SE-3400 contact probe surface profilometer. A part of the gel films was scraped off with a surgical knife before heat-treatment, and the level difference was measured after heat-treatment. Crack formation was examined by an optical microscope and by the contact probe surface profilometer. X-ray diffraction patterns were measured by a Rigaku RTP300 X-ray diffractometer with Cu Kα radiation operating at 50 kV and 150 mA. Polarization-electric field (P-E) characteristics of PZT films were evaluated at a frequency of 60 Hz using a Kenwood CS-4135 oscilloscope. For this measurement, the films were deposited on a nesa silica glass substrate and platinum electrodes of 0.2 mm diameter were deposited.

RESULTS AND DISCUSSION

BaTiO$_3$ Films

BaTiO$_3$ films were prepared from solutions containing various amounts of PVP and aged for various periods of time, where dip-coating was performed at a substrate withdrawal speed of 3 cm min^{-1}. The thickness of the fired films is plotted against sol aging time in Fig. 2. The crosses and circles in the figure denote the films with and without cracks, respectively. As seen in the figure the thickness increased with increasing sol aging time due to the increase in sol viscosity via polycondensation reaction, and the films finally cracked. The critical thickness lies between the crosses and circles in the figure, and it is seen that the critical thickness greatly increased over 1 μm when PVP was added to the starting solutions. The critical thickness evaluated appears to be almost constant at around 1.2 - 1.5 μm at

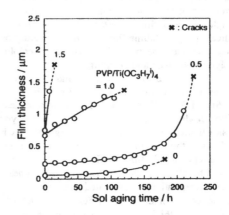

Fig. 2 Variation of the film thickness with sol aging time for the fired BaTiO$_3$ films prepared from sols containing various amounts of PVP via single-step dip-coating at a substrate withdrawal speed of 3 cm min^{-1}. Crosses and circles denote the films with and without cracks, respectively.

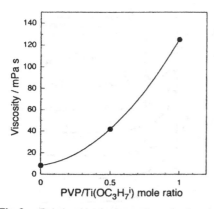
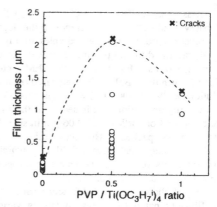

Fig. 3 Relationship between the sol viscosity and PVP/Ti(OC$_3$H$_7^i$)$_4$ mole ratio for the BaTiO$_3$ precursor sols aged for 15 min.

Fig. 4 Film thickness plotted against PVP/Ti(OC$_3$H$_7^i$)$_4$ mole ratio for the fired BaTiO$_3$ films prepared from sols shown in Fig. 3. Crosses and circles denote the films with and without cracks, respectively.

PVP/Ti(OC$_3$H$_7^i$)$_4$ = 0.5 to 1.5.

In another experiment, the critical thickness of the BaTiO$_3$ films were evaluated by varying the film thickness through changing substrate withdrawal (3.0 - 7.4 cm min^{-1}) or rotation (315 - 3440 rpm) speeds in dip- and spin-coating, respectively. The sols aged for 15 min were employed, and their viscosity is plotted against PVP/ Ti(OC$_3$H$_7^i$)$_4$ mole ratio in Fig. 3. The viscosity increased from 8 to 125 mPa s when the PVP/Ti(OC$_3$H$_7^i$)$_4$ ratio increased from 0 to 1. The thickness of the fired films is plotted against PVP/Ti(OC$_3$H$_7^i$)$_4$ mole ratio in Fig. 4, where different thickness at the same PVP/Ti(OC$_3$H$_7^i$)$_4$ ratio was achieved by changing the substrate withdrawal or rotation speeds. It is seen that the maximum thickness achieved without cracks increased from 0.2 to 2.1 μm when PVP/ Ti(OC$_3$H$_7^i$)$_4$ was increased from 0 to 0.5, and then decreased to 1.2 μm at PVP/Ti(OC$_3$H$_7^i$)$_4$ = 1.0. Difference is found in Fig. 2 and 4 on the dependence of the critical

Fig. 5 SEM picture of the 2.1 μm thick BaTiO$_3$ film prepared from the solution of PVP/Ti(OC$_3$H$_7^i$)$_4$ = 0.5 via single-step spin-coating and firing.

thickness on PVP/Ti(OC$_3$H$_7^i$)$_4$ ratio, which probably results from the difference in the degree of polycondensation between samples in Fig. 2 and 4, causing different polymeric structure of the hybrid gel films. Fig. 5 shows the SEM picture of the 2.1 μm thick BaTiO$_3$ film prepared from the solution of PVP/Ti(OC$_3$H$_7^i$)$_4$ = 0.5 via single-step spin-coating and firing. It is seen that the films are composed of particles ca. 0.06 μm in size, not very densely packed. However, although the films were converted from gel films containing large amounts of organic polymers, macropores are not detected in the film.

Fig. 6 shows the X-ray diffraction patterns of the films prepared from sols containing various amounts of PVP. The sols were aged for 15 min, and dip-coating was made at 3 cm min^{-1}. It is seen that single-phase BaTiO$_3$ is formed when PVP/Ti(OC$_3$H$_7^i$)$_4$ ≥ 0.4, whereas TiO$_2$ (rutile) is precipitated without BaTiO$_3$ when PVP/Ti(OC$_3$H$_7^i$)$_4$ = 0.2, indicating that PVP promotes the formation of the complex oxide, BaTiO$_3$. Because of the small film thickness no diffraction peak was observed PVP/Ti(OC$_3$H$_7^i$)$_4$ = 0

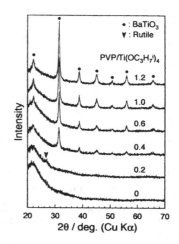

Fig. 6 X-ray diffraction patterns of the fired films prepared from the BaTiO$_3$ precursor sols containing various amounts of PVP. The sols were aged for 15 min.

PZT Films

Fig. 7 shows the thickness of the fired PZT films as a function of PVP/alkoxides ratio x in the

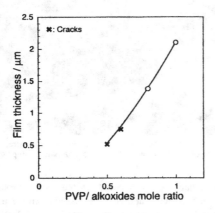

Fig. 7 Dependence of the film thickness on PZT/alkoxides mole ratio for the fired PZT films prepared from solutions of $y = 5$ via single-step dip-coating at a substrate withdrawal speed of 2 cm min^{-1}.

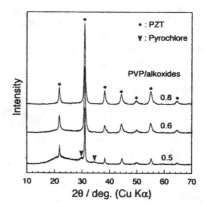

Fig. 8 X-ray diffraction patterns of the films prepared from PZT precursor sols containing various amounts of PVP where $y = 5$ via single-step dip-coating at a substrate withdrawal speed of 2 cm min^{-1}.

starting solutions where $y = 5$ and dip-coating was performed at a substrate withdrawal speed of 2 cm min^{-1}. The thickness increased linearly with increasing PZT content due to the increase in sol viscosity. It should be pointed out that crack formation occurred on firing when PZT/alkoxides ratio x was small at 0.5 and 0.6, whereas no cracks were observed in spite of the larger thickness when x was large at 0.8 and 1.0, showing the suppression of cracks by PVP. Fig. 8 shows the X-ray diffraction patterns of the films prepared from solutions of PVP/alkoxides =0.5 to 0.8 with $y = 5$. It is seen in Fig. 8 that single-phase PZT is formed when PVP/alkoxides = 0.6 and 0.8 whereas a trace amount of pyrochlore is observed when PVP/alkoxides = 0.5. Similarly to the case of BaTiO$_3$ described above, PVP promoted the formation of the complex oxide, PZT. The C=O group of PVP can coordinate metal ions as

Fig. 9 SEM picture of the 1.7 mm thick PZT film prepared via single-step deposition from the sol of $x = 0.9$ and $y = 10$.

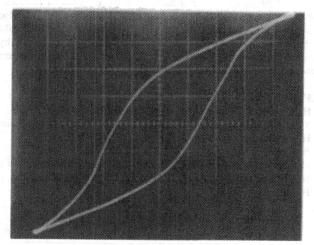

Fig. 10 P-E hysteresis loop of the PZT film shown in Fig. 9. P: 3.18 µC cm^{-2}/div., E: 6.33 kV cm^{-1}/div.

well as can make strong hydrogen bond with the OH groups of the metalloxane polymers, which makes the distribution of the metal ion species homogeneous, possibly leading to enhancement of the complex oxide formation. When PVP/alkoxides was smaller than 0.5, precipitation was formed in the solutions, indicating that PVP stabilizes the alkoxides against rapid hydrolysis.

A crack-free, 1.7 µm thick PZT film was prepared successfully via single-step deposition from a solution of $x = 0.9$ and $y = 10$. The film looked yellow and opalescent, transmitting the visible light. Fig. 9 shows the SEM picture of the film, where porous microstructure is observed in the cross-section with an denser overlayer with pinholes. The particle size observed in the picture was about 0.1 μm or less, while the crystallite size determined from the X-ray diffraction peak width was ca. 12 nm. In order to achieve dener films, further studies should be made so that the solution composition and the heat-treatment conditions be optimized. The 1.7 μm thick PZT film exhibited P-E hysteresis loop as shown in Fig. 10, which reveals the ferroelectricity of the film. Remanent poralization and coercive field of 5.7 μC cm^{-2} and 89 kV cm^{-1} were observed, respectively.

CONCLUSIONS

BaTiO$_3$ and PZT coating films were prepared by dip- or spin-coating from metal alkoxide solutions containing PVP.
(1) The critical thickness significantly increased by adding PVP to the starting solutions. Crack-free BaTiO$_3$ and PZT films 2.1 and 1.7 μm in thickness were obtained, respectively, via single-step deposition. The films, however, were not very dense, especially in PZT films.
(2) PVP in the starting solutions was found to enhance the formation of single-phase BaTiO$_3$ and PZT.
(3) Ferroelectricity of the 1.7 μm thick PZT film was observed in the P-E hysteresis loop.

ACKNOWLEDGMENTS

H. Kozuka thanks Dr. Susumu Tamura, Department of Electronics, Kansai University, for providing us suggestions on P-E hysteresis measurements. He also thanks Nippon Sheet Glass Foundation for Materials Science and Engineering and Kinki-chiho Invention Center for their financial support. This work was also financially supported by the Institute of Industrial Technology, Kansai University, and the Kansai University Research Grants (Grant-in-Aid for Encouragement of Scientists, 2000).

REFERENCES

1. H. Schmidt, G. Rinn, R. Naβ and D. Sporn, *Mat. Res. Soc. Symp.*, **121**, 743 (1988).
2. Y.-L. Tu, M.L. Calzada, N.J. Phillips, and S.J. Milne, *J. Am. Ceram. Soc.*, **79**, 441 (1996).
3. H. Kozuka and M. Kajimura, *Chem. Lett.*, 1029 (1999).
4. H. Kozuka and M. Kajimura, *J. Am. Ceram. Soc.*, in press.
5. H. Kozuka, M. Kajimura, T. Hirano and K. Katayama, *J. Sol-Gel Sci. Techn.*, in press.
6. H. Kozuka, M. Kajimura, K. Katayama, Y. Isota and T. Hirano, paper presented at 1999 MRS Fall Meeting.

PREPARATION AND CHARACTERIZATION OF $Sr_{0.48}Ba_{0.52}Nb_2O_6$ CERAMICS FIBERS THROUGH SOL-GEL PROCESSING

Masahiro Toyoda* and Kyougo Shirono
Fukui National College of Technology, Geshi, Sabae, 916-8507 Japan
*Corresponding author. E-mail address: toyoda22@fukui-nct.ac.jp (M. Toyoda)

ABSTRACT

The sol-gel processing was applied to the fabrication of $SrBaNb_2O_6$ ceramics fibers. Sr methoxide, Ba methoxide, Nb pentaethoxide and ethylene glycol monoethyl ether as starting materials were selected and then reflux with stirring to form complex alkoxide. The hydrolysis and polycondensation of its alkoxide gave polymerized products, and as a result the viscosity of the solution increased, its suggested that linear polymer products were obtained. The $SrBaNb_2O_6$ Gel fibers were drawn from the viscous solution. Its gel fibers were crystallized into tungsten bronze phase at 800 °C. The heat-treated fibers were a few centimeters long and from 30 to 200 μm long in diameter. The dielectric constant was 450 at 400 Hz.

INTRODUCTION

$Sr_xBa_{1-x}Nb_2O_6$ (SBN) has a tetragonal tungsten bronze structure and they have large pyroelectric coefficients and piezoelectric and electro optic properties. SBN has been receiving great attention for applications in pyroelectric sensors, SAW filters and electro-optic devices.[1-4] In particular, $Sr_{0.48}Ba_{0.52}Nb_2O_6$ has an attractive effect on the piezoelectric effect, linear optoelectric and high refraction index. $Sr_{0.48}Ba_{0.52}Nb_2O_6$ is expected to be applied for a piezo type infrared sensor,[5] an elastic surface wave filter [6] and hologram devices.[7] Its ceramics is also important material because of its lead-free composition. $Sr_{0.48}Ba_{0.52}Nb_2O_6$ ceramics have been prepared by the solid reaction of oxide powders at 1200 to 1400 °C.[8-10] $Sr_{0.48}Ba_{0.52}Nb_2O_6$ ceramics using hot pressing to improve the optical properties was developed by Okazaki et. al.[8] However, their solid reactions is necessary high temperature to obtain the final products.

Chemical processing such as sol-gel processing was adopted to prepare $Sr_xBa_{1-x}Nb_2O_6$ thin films ceramics due to some advantages, low temperature processing, etc.[11,12] Recently, the demand for fiber shapes processing has increased because of the integrated device and improved functionality of sensor. Control of the S/Ba ratio is one of the key factors to optimize the properties of $SrBaNb_2O_6$ ceramics. However, the control of composition is usually difficult in CVD and sputtering. Sol-Gel processing has some advantages such as high purity, feasible control of chemical stoichiometry and fabricating of fibrous shape of ceramics.[13,14] In this paper, an attempt is made to fabricate the $SrBaNb_2O_6$ ceramics fibers through sol-gel processing and to measure the electric properties.

EXPERIMENTAL PROCEDURE

Fig. 1 shows the experimental procedure of preparation of complex alkoxide solution of SrBaNb precursor. Sr methoxide [$Sr(OCH_3)_2 1/2H_2O$], Ba methoxide [$Ba(OCH_3)_2$], Nb pentaethoxide [$Nb(OC_2H_5)_5$] as raw materials and ethylene glycol monoethyl ether as organic solvent were selected. Ethylene glycol monoethyl ether was dried over molecular sieves and distilled before use. Sr methoxide and Ba methoxide were dissolved in ethylene glycol monoethyl ether at around boiling point of ethylene glycol monoethyl ether, 125°C, with stirring for 48 h. And then, its solution was mixed with $Nb(OC_2H_5)_5$ and excess its organic solvent. It s mixed solution was reacted at ethylene glycol monoethyl ether, 125°C, with stirring for 48 h. Then, obtained precursor solution was concentrated and removed by-products by vacuum distillation. The final solution, which had a concentration 0.5 M was yellowish transparent solution without any suspension of particles. All procedures were conducted in dry N_2 atmosphere, because the starting materials are extremely sensitive to moisture.

For the investigation of the gelation, the 0.5 M Sr-Ba-Nb complex alkoxide solution was weighed and combined with 10 ml of ethylene glycol monoethyl ether based solution which contained a specific concentration water and additive, HCl, in a sample bottle. The gelation reaction was designated by the additive and the molar ratio of water to alkoxide.

The spinnability of the sol-solutions was examined from the capability of fiberization of the sols by dipping a glass stick and pulling it up. The obtained fibers were heat-treated at 700 and 800 °C.

The observation of morphology on fibers was conducted by using SEM (Hitachi, S-4100). The crystalline phase of its fibers was identified by XRD (Rigaku, Rad-2B).

The dielectric measurement was examined using Hewlett Packard impedance analyzer HP4284A.

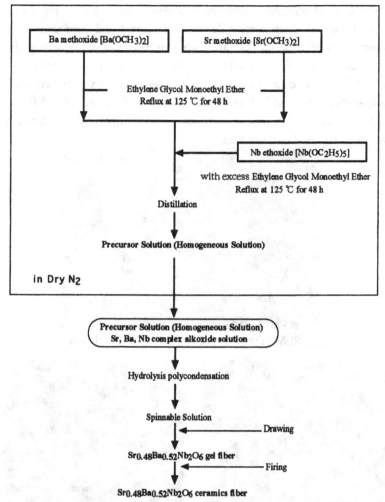

Fig. 1 Flow diagram of experimental procedure for preparation of complex alkoxide solution and drawing $Sr_{0.48}Ba_{0.52}Nb_2O_6$ ceramics fiber

RESULT and DISCUSSION

The viscosity of the Sr-Ba-Nb complex alkoxide solution increased with the passage of time by hydrolysis polycondensation. The viscosity of solution increased up to about 1 kg/m·sec, and then became slower in increase above this value by hydrolysis. The gel fibers were spinnabled through viscous sol-solutions by dipping a glass stick and pulling.

Fibers could be drawn after the viscosity reached about 1 kg/m·sec.   Fig. 2 indicates gel fiber.   Its drawn gel fibers were yellowish transparent and homogeneous.   Fig. 3 shows morphology of ceramics fiber after heat-treated at 800 ℃ [low magnification (a) and high magnification (b)].   The heat-treated fibers were a few centimeters long and from 30 to 200 μm in diameter, crack free and densed.   As seen, grains of its fibers were flake-shaped.   Size of flaked-shaped grain was about 0.2 × 0.5 μm.   Fig. 4 indicates XRD patterns of the fiber.   The Sr-Ba-Nb gel fibers heat-treated at 800 ℃ is proved to have the crystallized tungsten

Fig. 2 SEM micrograph of $Sr_{0.48}Ba_{0.52}Nb_2O_6$ gel fiber

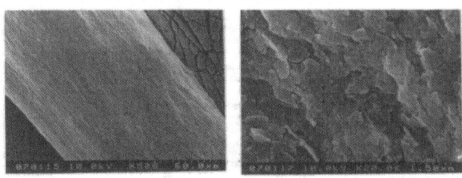

(a) low magnification (b) high magnification

Fig. 3 SEM micrograph of $Sr_{0.48}Ba_{0.52}Nb_2O_6$ ceramics fiber fired at 800 ℃

bronze phase.   Hirano and co-workers[11] reported that the formation of solid solution with potassium was found to be very effective in forming the tungsten bronze phase at lower temperatures at 700 ℃ without any formation of the orthorhombic low-temperature. However, tungsten bronze phase prepared by this processing was crystallized at 800 ℃

without potassium. Molar ratio of Sr-Ba-Nb complex alkoxide solution and composition of $Sr_{0.48}Ba_{0.52}Nb_2O_6$ ceramics fiber analyzed close to the expected stoichiometry (Sr/Ba = 0.465/0.535).

To measure the ferroelectric properties, both ends of fiber, 2 cm long and 100 μm in diameter, was fixed on glass substrate by using Ag paste. The dielectric constant and dielectric loss of its fiber fired at 800 °C were 450 and 2.3 %, respectively at room temperature and 400 Hz.

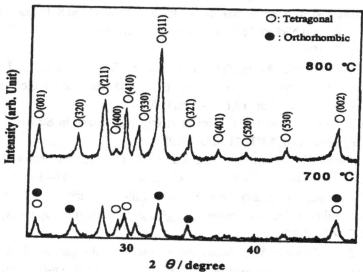

Fig. 4 XRD patterns of $Sr_{0.48}Ba_{0.52}Nb_2O_6$ ceramics fiber fired at 700 and 800 °C

CONCLUSION

1. $Sr_{0.48}Ba_{0.52}Nb_2O_6$ ceramics fiber was successfully prepared through sol-gel processing.
2. Sr-Ba-Nb complex alkoxide solution produce tungsten bronze phase was obtained heat-treatment at 800 °C.
3. Grains of $Sr_{0.48}Ba_{0.52}Nb_2O_6$ fibers were flake-shaped. Size of these flaked-shaped grain were about 0.2 × 0.5 μm.
4. The dielectric constant and dielectric loss of $Sr_{0.48}Ba_{0.52}Nb_2O_6$ ceramic fiber fired at 800 °C were 450 and 2.3 % at room temperature at 400 Hz, respectively.

REFERENCES

[1] M. P. Trubeja, E. Ryba and D. K. Smith, "A Study of Positional Disorder in Strontium Bariumu Niobate," J. Mater. Sci., **31**, 1435-1443(1996).

[2] R. R. Neurgaonkar, W. F. Hall, J. R. Olover, W. W. Ho and W. K. Cory, Tungsten Bronze $Sr_{1-x}Ba_xNb_2O_6$: A Case History of Versatility." Ferroelectrics, **87**,167-179(1988)

[3] A. M. Glass, "Investigation of the Electrical Properties of $Sr_{1-x}Ba_xNb_2O_6$ with Special Reference to Pyroelectric Detection," J. App;. Phy., **40**[12],4699-4713(1969)

[4] D. Ryta, B. A. Wechsler, R. N. Schwartz, C. C. Nelson, C. D. Brandle, A. J. Valentino and G. W. Berkstresser, "Temperature Dependence of Photorefractive Properties Strontium-Bariumu Niobate ($Sr_{0.6}Ba_{0.4}Nb_2O_6$)," J. Appl. Phys., **66**[5],1920-1924(1989)

[5] S. Nishiwaki, T. Yogo, K. Kikuta, K. Ogiso, A. Kawase and S. Hirano, "Synthesis of Strontium-Bariumu Niobate Thin Films Through Metal Alkoxide," J. Am Ceran, Soc., **79**[9], 2283-2288(1996)

[6] R. R. Neurgaonka, M. H. Kalisher, T. C. Cim, E. J. Staples and K. L. Keester, "Czochralski Single Crystal Growth of $Sr_{0.61}Ba_{0.39}Nb_2O_6$ for Surface Acoustic Wave Applications," Mater. Res. Bull., **15**,1235-1240(1980)

[7] J. B. Thaxter, "Electrical Control of Holographic Strage in Strontium-Bariumu Niobate," Appl. Phys. Lett., **15**(7), 210-212(1969)

[8] K. Nagata, Y. Yamamoto, H. Igarashi and K. Okazaki, "Properties of the Hot-Pressed Strontium-Bariumu Niobate Ceramics," Ferroelectrics, **38**, 853-856(1981)

[9] S. Nishiwaki, J. Takahashi and K. Kodaira, "Efffect of Additives on Microstructure Development and Ferroelectric Properties of $Sr_{0.3}Ba_{0.7}Nb_2O_6$ Ceramics," Jpn, J. Appl. Phys., **33**[9B], 5477-5481(1994)

[10] N. S. VanDamme, A. E. Sutherland, L. Jones, K. Bridger and S. R. Winzeer, "Fabrication of Optical Transparent and Electrooptic Strontium-Bariumu Niobate Ceramics," J. Am Ceran, Soc., **74**[8], 1785-1792(1991)

[11] W. Sakamoto, T. Yogo, A. Kawase and S. Hirano, "Chemical Processing of Potassium-Substituted Strontium Barium Niobate Thin Films through Metallo-Organics." J. Am. Cera. Soc., **81**[10], 2692-2698(1998)

[12] K. Nishio, N. Seki, J. Thongrueng, Y. Watanabe and T. Tsuchida, "Preparation and Properties of Highly Oriented $Sr_{0.3}Ba_{0.7}Nb_2O_6$ Thin films by a Sol-Gel Processing." J. Sol-Gel Sci. Tec., **16**, 37-45(1999)

[13] M. Toyoda, Y. Hamaji and K. Tomono, "Fabrication of $PbTiO_3$ Ceramics Fibers by Sol-Gel Processing." J. Sol-Gel Sci. Tec., **9**, 71-84(1997)

[14] M. Toyoda, Y. Hamaji and K. Tomono, "Fabrication of $Pb(Ti,Zr)O_3$ Ceramics Fibers by Sol-Gel Processing." Nippon Kagakukai-shi, **1995**; #2, 150-156(1995)

SOL GEL PROCESSING FOR THE LANTHANIDES

James L. Woodhead
AMR Technologies Inc.
121 King Street West, Suite 1740
Toronto, Ontario
M5H 3T9

ABSTRACT
Sol Gel processes have been developed to produce lanthanide-based materials with enhanced and pre-determined properties e.g. particle size and density. The processes use inorganic starting materials and key issues for scale-up are discussed.

INTRODUCTION
During the 1960s and 1970s very extensive research and development work was carried out in the Oakridge Laboratory USA,* and at the United Kingdom Atomic Energy Authority Establishment Harwell** on inorganic sol gel processes. This research work identified chemical procedures for de-aggregating hydrous metal oxides to give nano-sized primary particles that could reassembled into precursor sols and gels that could yield oxide products with predetermined properties such as particle shape, size, density and composition. The heat treatment procedures were very energy efficient yielding oxides with densities close to theoretical at <1000°C. Much of this early research work used lanthanide salts and hydrous oxides as simulates for the radioactive species, e.g. thorium oxide and uranium oxides.

There is now an increasing worldwide demand for lanthanide and lanthanide-based materials with strict specifications on composition, particle size and shape, and product density and pore structure. To meet these requirements, the underlying science and technology base established in earlier years for inorganic sol gel processes has been exploited and further developed to meet the needs of rare earth users.

---

*E. Ferguson, O.C. Dean and D.A. Douglas; *3rd Introduction on the Peaceful Uses of Atomic Energy*, Geneva 1964 A/Conf. 28 P237
**E.S. Lane, J.M. Fletcher, J.M. Holdaway, K.R. Hyde, C.E. Lyon and J.L. Woodhead AERE R-5241 (1966)

To the extent authorized under the laws of the United States of America, all copyright interests in this publication are the property of The American Ceramic Society. Any duplication, reproduction, or republication of this publication or any part thereof, without the express written consent of The American Ceramic Society or fee paid to the Copyright Clearance Center, is prohibited.

Freshly pptd., aged 18 min.

Ppt. washed, aged 80 min.

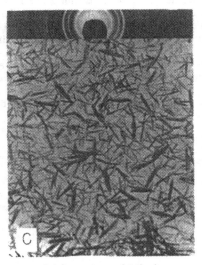
Sol fom ppt., aged 140 min.

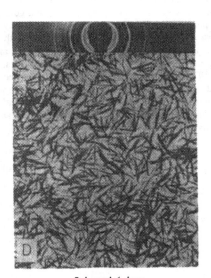
Sol aged 6 days

SCALE ⊢⊣ = 600 Å

AERE - R 5619 Fig. 2
Electron micrographs and electron diffraction patterns of
praseodymium hydroxide precipitates and sols

EXPERIMENTAL

Lanthanide hydroxides were prepared by dissolving nitrate salts in water and precipitating with tetra-methyl ammonium hydroxide, ammonium hydroxide or sodium hydroxide. After extensive washing with water, the hydroxide gels were peptized with nitric acid.

Cerium IV hydrate was prepared by oxidising CeIII chloride solutions with sodium hypochlorite. After washing with dilute ammonium hydroxide, the Cerium IV hydrate was de-aggregated using dilute nitric acid.

RESULTS AND DISCUSSION

The Process of De-aggregation

Fig. 1 shows the changes that occur when freshly precipitated $Ln(OH)_3$ is washed free from entrained ions and then aged (photo-micrographs A and B); the further changes that occur when the hydrous oxide is peptized with dilute mineral acid is shown in photo-micrographs C and D. The de-aggregation and peptizing steps results in the formation of an almost transparent colloidal dispersion that can be evaporated to a water-dispersible gel. When heated in air, the gel is rapidly transformed into a dense oxide with a fine grain structure.

Cerium IV hydrate can be de-aggregated with dilute mineral acid to give 5 to 10 nm particles which can be dispersed in water to form sols containing up to 450g/L of cerium oxide equivalent. X-ray diffraction shows that no changes in the crystallite size occurs as a result of the de-aggregation.

The Use of Sol Gel Processes on an Industrial Scale

The use of inorganic starting materials is of great benefit for the scale-up of sol gel processes. Organo metallic reagents have been widely used for the preparation of gel-processed materials e.g. coatings and nano-sized particulates. However, the starting materials can be very expensive, and on a large scale the required hydrolytic reactions can be difficult to control.

Hydrous oxides are readily available from within the chemical processing industries and an understanding of the de-aggregation step should enable relatively inexpensive starting materials to be converted into colloidal dispersions suitable for sol-gel processing. Processes for alumina, zirconia and ceria have found wide scale application, and have been operated on a multi-tonne scale.

Sol-Gel Processes for Cerium Oxides

British Patent 1,342,893(3) describes the preparation of ceria sol using a ceria hydrate starting material and aqueous nitric acid as the de-aggregating agent. The carefully characterized ceria hydrate was slurried with nitric acid to give a slurry with a pre-determined HNO_3/CeO_2 mole ratio, the slurry was then digested at an elevated temperature (40 – 90°C) to produce the required degree of de-aggregation.

Complete de-aggregation produces a conditioned hydrate that is fully dispersible in water (sol). However, carefully pre-calculated degrees of de-aggregation was shown to produce gels and oxide products with a range of densities. Oxide powder densities are usually controlled by calcination temperature or by using different salts, e.g. oxalate or nitrates. The de-aggregation process described here provides a method of density control using the same starting material and the same calcination temperature. The latter feature exercises control over crystallite size. Table I shows typical results.

Table I. The effect of nitric acid concentration on the de-aggregation of cerium IV hydrate

HNO_3/CeO_2 mole ratio	Gel density (100°C), g/cc	Oxide density (1000°C), g/cc
0	1.00	1.15
0.25	1.20	1.50
0.53	1.80	2.80
1.00	2.00	3.10

CONCLUSION

Hydrous oxides of the lanthanides derived from inorganic starting materials can be converted to colloidal intermediates which are suitable for further processing to give oxide products with pre-determined properties. Such processes have been operated on a large scale and there are now opportunities to use such processes to prepare lanthanide and lanthanide-based materials with superior properties.

SOL-GEL DERIVED NICKEL TITANATE FOR TRIBOLOGICAL COATINGS

D.J. Taylor and P.F. Fleig
TPL, Incorporated
3921 Academy Parkway North
Albuquerque, NM 87109

R.A. Page
Mechanical & Materials Engineering Division
Southwest Research Institute
San Antonio, TX 78228

ABSTRACT

It was previously discovered that titanium and nickel ions implanted into silicon nitride produced good tribological surfaces. Analysis of these surfaces after they had experienced mechanical mixing (friction and wear tests) revealed mixed titanium/nickel oxide. We attempted to duplicate these lubricious surfaces *a priori* by sol-gel deposition of titanium oxide and nickel oxide precursors. During this study, we determined that titania-rich nickel titanate compositions yielded the best tribological performance in the TiO_2-NiO system. The microstructure of these compositions was characterized by nickel titanate crystals and rutile, both nominally 50 nm in diameter. Coatings were uniform and flat and had excellent adhesion to various substrates. The coefficient of friction between tested materials was reduce by at least 50% and the coatings showed little to no wear.

INTRODUCTION

A previous investigation of surface-modified silicon nitride showed a significant decrease in friction and wear.[1] In this study, silicon nitride monoliths were modified with various metal atoms using ion implantation. Combinations of titanium and nickel implanted into the surface of silicon nitride yielded a coefficient of friction (COF) of 0.09 at 800°C against a titanium carbide (TiC) pin. Examination of the surface layers by Auger spectroscopy indicated that the low friction coefficients were obtained on surfaces with a titanium oxide film that had substantial nickel content.

In the present work, we developed wet chemical methods of depositing the titanium oxide materials believed to be responsible for the improved tribological performance observed in the previous study. We hypothesized that deposition of the oxides as nanostructured materials would promote lubricious behavior over a range of temperatures. Microstructure was examined by X-ray diffraction (XRD), energy dispersive spectroscopy (EDS) and scanning electron microscope (SEM). The COF

To the extent authorized under the laws of the United States of America, all copyright interests in this publication are the property of The American Ceramic Society. Any duplication, reproduction, or republication of this publication or any part thereof, without the express written consent of The American Ceramic Society or fee paid to the Copyright Clearance Center, is prohibited.

was measured by reciprocating ball-on-plate at 300 K and 773 K. Wear tracks were observed with surface profilometry, optical microscopy and SEM. Titania/nickel oxide coatings reduced the COF by at least 50% at both temperatures and demonstrated the best wear resistance.

The goal of this study is to develop sol-gel derived coatings as solid lubricants over a wide temperature range. The coatings will be used to reduce friction and wear in engine and bearing applications. The sol-gel approach facilitates the application of a wide variety of ceramic coatings to most substrate materials or geometries. Several compositions are being investigated. Tribological coatings will be characterized for friction/wear applications. In this paper, we present results obtained during this on-going study with respect to processing parameters and microstructure.

EXPERIMENTAL

Coating Preparation

Coating solutions were synthesized from titanium (IV) isopropoxide and nickel (II) acetate tetrahydrate. The amount of each precursor was calculated based on the final oxide composition (in weight percentage). Enough ethanol was used to produce a 20:1 molar ratio with the combined precursors. Nickel (II) acetate tetrahydrate was dissolved in distilled ethanol and stirred as concentrated nitric acid was added. After several hours, titanium (IV) isopropoxide was added slowly by syringe. The solutions were allowed to stir in sealed vessels for several hours before coating.

Films were deposited onto cleaned and polished Hastalloy X substrates by spin coating and dip coating. Deposition conditions were 2000 rpm for 30 sec, and 4 mm/s withdrawal rate, respectively. Coated coupons were fired to 1073 K. Witness samples were made on silicon by both methods for easy analysis with ellipsometry.

After coating, the remaining solution was allowed to gel. The gel was dried, crushed by mortar and pestle, and fired along with the coated coupons. These powders were used for X-ray diffraction analysis. Heat treatments were performed on the powders at higher temperatures and over longer durations to determine if crystal growth would be problematic.

Coating Evaluation

Coating thickness and refractive index was measured using multi-angle ellipsometry (Sentech SE400-11) at 6328 Å wavelength. Microstructure was examined by XRD (Siemens D-500) using the powders obtained from the dried gels.

Friction and wear were measured on an oscillatory pin-on-flat apparatus.[2] The tests were performed at 300 K and 773 K in air, using a 1 N load at 2 cm/s. The stroke of the tester was 1 cm, which corresponded to a 1 Hz oscillation frequency. Wear tracks were examined with optical microscopy, optical profilometry (Wyko RST Plus), SEM (Amray 1645) and EDS (Tracor Northern Micro-Z II).

RESULTS

Ellipsometry was employed to determine thickness and index of refraction of the coatings. Individual layer thickness depended upon each precursor solution and the deposition conditions. When thickness was below 0.5 micron, multiple layers were deposited to achieve coatings that were at least this thick. Refractive index (RI) varied with composition. For example, RI of the 75% TiO_2 - NiO composition fired to 1073 K was 2.06, which indicated 29% porosity based on the theoretical RI (2.5) for this composition. Porosity was calculated using literature values, experimental values and the Maxwell Garnett effective medium approximation model. This model assumes spherical inclusions (pores in this case) and dipole interactions, which are standard first approximations that give reasonably accurate results.[3]

X-ray diffraction spectra showed the crystalline phases present in each composition. Crystallite size was calculated from the diffraction peaks using the Sherrer formula.[4] The XRD results of several TiO_2-NiO compositions are given in Table I. The titanium oxide and nickel oxide precursors were combined in such a way as to produce a mixed metal oxide phase, in this case, nickel titanate ($NiTiO_3$). Figure 1 displays an XRD spectrum for 50% TiO_2 - 50% NiO showing the formation of nickel titanate. Crystallite sizes for all phases were consistently between 40 nm and 50 nm and remained virtually unchanged after being fired to 1073 K, 1273 K, 1473 K or 1673 K for one hour, or at 1073 K for 8 hours.

Table I: Crystallinity of Wet-Chemically Derived Nickel Titanates

Composition (in weight percent)	Phases	Crystallite Size (nm)
25% TiO_2 - 75% NiO	$NiTiO_3$	44
	NiO	44
33% TiO_2 - 67% NiO	$NiTiO_3$	49
	NiO	45
50% TiO_2 - 50% NiO	$NiTiO_3$	47
	TiO_2 (rutile)	47
	NiO	51
67% TiO_2 - 33% NiO	$NiTiO_3$	42
	TiO_2 (rutile)	40
	NiO	48
75% TiO_2 - 25% NiO	$NiTiO_3$	48
	TiO_2 (rutile)	44

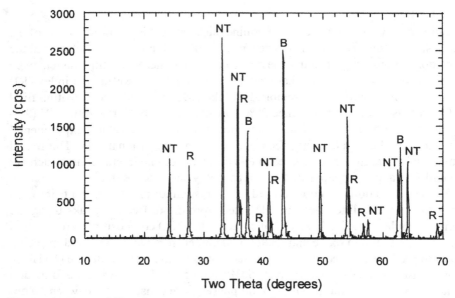

Figure 1: X-ray diffraction spectrum of 50% TiO_2-NiO made by sol-gel methods. NT = nickel titanate ($NiTiO_3$), R = Rutile (TiO_2), B = Bunsenite (NiO)

Coefficient of friction values ranged from 0.08 to over 0.8, with 0.9 being the COF for a TiC pin on the Hastalloy substrate. Figures 2 and 3 show COF as a function of time in the friction tester at 300 K and 773 K, respectively. The material with the best performance at both temperatures was the titania-rich TiO_2-NiO compositions. The 67% and 75% TiO_2 compositions (balance nickel oxide) had similar performance, and both had lower COFs than the 50% TiO_2 composition. Nickel-rich compositions also performed better than the 50% TiO_2 composition, but not as well as those rich in titania.

Optical profilometry was employed to evaluate wear tracks that resulted after 3600 cycles on the friction tester. All of the TiO_2-NiO compositions exhibited excellent wear resistance; specifically, no wear tracks were evident after one hour of testing (3600 cycles). To confirm the presence of the coating, we analyzed the surface with EDS in the SEM. For TiO_2-NiO coatings, the spectra were identical inside and outside of the wear track. Since nickel was common to both coating and substrate (Hastalloy X contains 49% Ni, 22% Cr, 20% Fe and 9% Mo), it was the titanium and oxygen signals that indicated the presence of the coating. Concerning the counterfaces, optical microscopy showed little wear on the TiC pins.

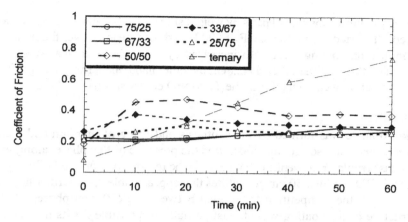

Figure 2: Coefficient of sliding friction as a function of time for sol-gel derived nickel titanate compositions on Hastalloy at 300 K (TiC pin). The binary compositions are listed by weight percent TiO_2/NiO and the ternary composition is 33% each of TiO_2, SiO_2 and NiO.

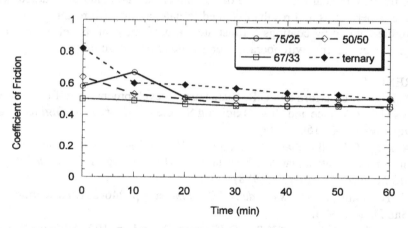

Figure 3: Coefficient of sliding friction as a function of time for sol-gel derived nickel titanate compositions on Hastalloy at 773 K (TiC pin). Legend designations are the same as in Figure 2.

A ternary composition (equal parts by weight silica, titania, and nickel oxide) was also tested. It started with a low COF at 300 K and rose steadily over the hour test period, indicating coating wear with increased cycles. The wear scar was more pronounced on this film than the others. On the other hand, the ternary composition had similar friction/wear behavior to the TiO_2-NiO compositions at 773 K.

DISCUSSION

Nickel titanate was formed by the synthesis and firing of titanium and nickel oxide precursors using sol-gel methods. It is the presence of this phase, along with a minor phase (rutile or bunsenite), that seems to be responsible for good tribological properties. XRD revealed 50 nm crystallites that appear stable upon further heating, probably due to the competition for growth between the different phases. If the microstructure of the coating was at least partially responsible for its tribological properties, maintaining the microstructure is crucial.

The nickel titanate coatings significantly decreased the COF between Hastalloy and TiC and exhibited good wear properties. Tests performed at elevated temperatures were equally successful. Further experiments include depositing nickel titanate onto silicon nitride, which will allow us to increase the firing temperature of the coatings, thereby increasing their density. Less porous coatings should further increase the tribological performance of the nickel titanate materials studied here. All of the experiments and results obtained to date, along with results from other nickel oxide-containing coatings,[5] illustrate the usefulness of nickel titanate as a tribological material that will operate over a wide temperature range.

REFERENCES

[1] R.A. Page, J. Lankford and C.R. Blanchard, "Development of Self Lubricating Ceramics Using Surface and Bulk Oxidizing Species," *Advances in Engineering Tribology*, **SP-31** 145-150 (1990).

[2] R.A. Page, C.R. Blanchard-Ardid and W. Wei, "Effect of Particulate Additions on the Contact Damage Resistance of Hot-Pressed Si_3N_4," *Journal of Materials Science*, **23** 946-957 (1988).

[3] H.G. Tompkins, *A User's Guide to Ellipsometry*, pp. 246-251. Academic Press, San Diego, 1993.

[4] B.D. Cullity, *Elements of X-Ray Diffraction*, 2nd ed., p. 102. Addison-Wesley Publishing Company, Inc., Reading, Massachusetts, 1978.

[5] Ph. Colomban, S. Jullian, M. Parlier and P. Monge-Cadet, "Idenification of the high-temperature impact/friction of aeroengine blades and cases by micro Raman spectroscopy," *Aerospace Science and Technology*, **3** [7] 447-459 (1999).

SYNTHESIS OF WOOD-(SiO_2,Al_2O_3) INORGANIC COMPOSITES BY SOL-GEL PROCESS

Xicheng Wang, Shulan Shi, Hongmei Hou, Peng Shi and Li Sun
Beijing Graduate School, Wuhan University of Technology
Beijing 100024, CHINA

ABSTRACT

A novel wood-(SiO_2,Al_2O_3) inorganic composite was developed by sol-gel process. The (SiO_2,Al_2O_3) ceramic precursor was made from reaction products of the TEOS, GPTMS and $Al_2(OH)_nCl_{6-n}$ chemical system. It was impregnated into the wood cell walls and *in situ* converted to (SiO_2,Al_2O_3) gels. As a result of this process, properties of the wood-(SiO_2,Al_2O_3) inorganic composites were distinctly enhanced.

INTRODUCTION

Wood is a super natural composite material. However, its characteristic features of flammability, biodeterioration and dimensional instability often restrict utilization of the wood. A number of attempts have been made to overcome such disadvantageous characteristics,. One of the current developments is the application of sol-gel process to prepare wood-inorganic composites.[1,2] A basic principle of this process involves a hydrolysis of silicon alkoxide solution with water and subsequent *in situ* polycondensation to produce SiO_2 gels in cell walls of the wood. However, the conventional sol-gel method for wood-inorganic composite was itself incomplete. First, since speeds of the hydrolysis and condensation reactions slow down with the lowering of H^+ concentration, the whole process takes a long period of time (5-6 days). Second, the weight percent gain (WPG) of the composites will decrease slowly. Thus, some improved properties of the composites will deteriorate as the time goes by.

In the present paper, we provided a modified sol-gel process with a highly basicity basic aluminium chloride(AS). In this process, AS was utilized as a replacement of HCl to catalyze the hydrolysis and condensation of TEOS and as a modifier component to yield a substantially (SiO_2,Al_2O_3) oxide network.

EXPERIMENTAL

Preparation of AS Solution

Preparation of the AS was carried out by the starting material aluminium chloride hexahydrate ($AlCl_3 \cdot 6H_2O$). It was heated at $130\,^\circ C$ to get basic aluminium chloride. Then, it was dissolved in distilled water and treated by ionic exchange resin to produce highly basicity basic aluminium chloride $Al_2(OH)_nCl_{6-n}$(AS) with B>90% and B=(n/2×3)×100%.

Preparation and Test of Wood-inorganic Composites

Two kinds of precursors were made for wood-inorganic composites producing. Precursor A was prepared with TEOS/ethanol/GPTMS/ HCl/H$_2$O, while precursor B with TEOS/ethanol/ GPTMS/AS.

The precursor A and B were impregnated respectively into wood specimens (20×20×30mm^3) obtained from the sapwood portion of T.a Ruper. The impregnated specimens with precursor B were then placed for 12h in a oven temperature at 60°C and for 24h at 105°C. But the impregnated specimens with precursor A were treated following our previous process.[2]

By measuring the oven-dry weight of the composite specimens, the WPG of the composites were determined. The oven-dry specimens were tested by SEM-EDXA technique with S4200 to determine the distribution of Si and Al elements within the composite. The thermal properties of the composite specimens were studied with a thermogravimetric analyzer Dupont1090. For testing antishrink efficiency(ASE), the specimens were soaked in distilled water for 3 days, then the changes in volume of the specimens were measured.

RESULTS AND DISCUSSION
Mechanism of the Reactions with AS and TEOS

The sol-gel method provided intimate mixing of aluminium and silica and oxide bond formation at low temperature. In the case of wood-inorganic composite process, however, aluminium tri-*sec*-butoxide is so reactive with water that the pores in the wood appeared to be blocked before the impregnation was complete. One thus could never get (SiO$_2$,Al$_2$O$_3$) oxides network within the cell walls of the wood when the aluminium alkoxides were used as the starting-materials.

In this study, it's considered that the AS replaced HCl firstly as a catalyst, and secondly as a modifier to build the structure of the (SiO$_2$,Al$_2$O$_3$) oxides network. During the reaction processes, it was observed that the pH in precursor B increased from 4.8 to 5.1 when the AS (pH=5.4) was added. There was little change at this pH value during a period of 1.5~3 hours. Then the pH would be lowering from 5.1 to 4.8, and subsequently to 4.1 in half an hour. The mechanism of hydrolysis-condensation reactions were presumed as follow:

Although H$^+$ concentration was lower in the precursor B, the concentration of Cl$^-$ released from AS was relatively higher (0.0139mol/l), which takes part in the catalytic reaction.[3] It would be able to promote the hydrolysis of TEOS. Simply shown:

$\equiv$Si-OC$_2$H$_5$ + H$^+$ +Cl$^-$ → $\equiv$Si-Cl + C$_2$H$_5$OH   (1)
$\equiv$Si-Cl + H$_2$O → $\equiv$Si-OH + H$^+$ +Cl$^-$ (2)
$\equiv$Si-Cl + HO-Si$\equiv$ → $\equiv$Si-O-Si$\equiv$ + H$^+$ +Cl$^-$ (3)

For the AS, its hydrolysis reaction (at 4<pH<6) could be shown as follow:[4]

[Al$_8$(OH)$_{22}$(H$_2$O)$_5$]$^{2+}$ → [Al$_8$(OH)$_{23}$(H$_2$O)$_4$]$^+$ +H$^+$ (4)

As said above, there was no immediate change in the pH of the precursor B with AS added; however the pH did lower after 1.5~3.0 hours. This should be due to the H$^+$ from the reaction (4) took part in the TEOS hydrolysis process. After this, the reaction (4) will go a step further and result into the H$^+$ concentration increasing. If the aluminium from AS would be not reacted with TEOS, the reaction (4) should shift to left direction and the value of pH should increase. But our experiments have shown that the pH of the system has not any increasing during the whole reaction processes. That indicate the aluminium from AS also takes part in the condensation of TEOS. Simply shown:

$$\equiv Al-OH + HO-Si\equiv \;\rightarrow\; \equiv Al-O-Si\equiv \;+\; H_2O \qquad (5)$$

$$\equiv Al-OH + Cl-Si\equiv \;\rightarrow\; \equiv Al-O-Si\equiv \;+\; H^+ + Cl^- \qquad (6)$$

Our experiments have indicated that the pH decreasing of the precursor B during the sol-gel process can shorten the gelling time from 5~6 days to 12~24 hours. Additional evidence for the presence of Si-O-Al linkages was demonstrated by IR spectroscopy (Fig.1). It can be seen from the IR spectra of SiO_2 and (SiO_2,Al_2O_3) gels. That Si-O-Si stretching vibration by SiO_2 gels gives a band at 1082 cm^{-1}, while the stretching vibration by (SiO_2,Al_2O_3) gel at 1072 cm^{-1}. It was shifted down to lower bands. Meanwhile, the O-Si-O stretching vibrations by SiO_2 gel and (SiO_2,Al_2O_3) gel give two bands at 949 cm^{-1} and 930 cm^{-1} respectively. It was also shifted down to lower bands. As have known, Al-O bond distance was longer than Si-O, and its bond energy was lower. The proportion of bond-force-constants K of Al-O and Si-O-Al to vibration frequency was smaller than that of the Si-O and Si-O-Si bonds. On the basis of these facts, the appearance of the vibration frequency being shifted to lower bands suggests the presence of Al-O-Si linkages in products (SiO_2,Al_2O_3) gels.

Distribution of (SiO_2,Al_2O_3) Gels in Wood

Figure 2 showed a SEM micrograph and the distribution maps of Si-K_α X-rays and Al-K_α X-rays in the cross section of the wood-(SiO_2,Al_2O_3)inorganic composite. It should be noted that Si-K_α X-rays and Al-K_α X-rays are emitted only from the cell walls of the composites. Thus, it may be concluded that the (SiO_2,Al_2O_3)gel has been formed within the cell walls of the composites. On the other hand, it has been proved in our previous studies[5] that in the TEOS and 3-glycidylpropyltrimethoxysilane(GPTMS) system, silica gel could be chemically bonded with cell wall through GPTMS. In the case of TEOS, GPTMS and AS system, it had been also evidenced by Shi's study [6] that the (SiO_2,Al_2O_3)inorganic gels can be also chemically bonded with wood through GPTMS. Simply shown:

$$-O-CH-CH_2-O-CH_2-CH_2-CH_2-Si-O-Si-O-Al-OH-Al-OH-Al\equiv$$
$$H_2C-O-wood$$
(with OH groups on Al)

Properties Of Wood-(SiO_2,Al_2O_3)Inorganic Composites
Dimensional Stability

As have known, the poor dimensional stability of wood was due to there were the active hydroxyl groups in it. In the wood-inorganic composite, a number of hydroxyl groups have been reacted with the gels, and furthermore, the reaction was carried out more completely in the wood-(SiO_2,Al_2O_3)inorganic composite. Thus, the wood-(SiO_2,Al_2O_3)inorganic composite shows lower swellibility.

Fig.3 showed the different dimensional stability of both composites prepared from precursor A and B at a certain WPG level. The composite B showed a higher ASE.

As Shi's study[6] has reported that AS not only catalyzed the hydrolysis of TEOS, but also hydrolysis of GPTMS and the condensation between GPTMS and the hydroxyl groups in the cell walls of the wood. So, the gels in the composite B have been bonded chemically with more hydroxyl groups than that in the composite A. It would result in that composite B has a higher dimensional stability.

Flammability

Fire tests were performed on samples: composite A, composite B and untreated wood. The results were shown in Table 1. Although composite A and B have the same WPG level, their degradation rate are different. It can be seen that the (SiO_2,Al_2O_3)gel has provided more effective isolation between wood and flame and oxygen, which results in that the composite B has better fire-resisting properties.

CONCLUSION
1. The highly basicity basic aluminium chloride(AS) can be utilized as catalysis and modifier for the sol-gel process of wood-(SiO_2,Al_2O_3)inorganic composites.
2. The wood-(SiO_2,Al_2O_3)inorganic composites showed the enhanced properties with the dimensional stability and flammability.

ACKNOLOWLEDGEMENT

This research was supported by the NSFC (No.59572041) and BSFC (No.2952009).

REFERENCES

[1] Shiro Saka and Fhiro Tanno, *Mokugai Gakkaish*i , Vol. 42 (1) 81-86 (1996)

[2] Xicheng Wang and Jie Tian, *Chinese Journal of Material Research* Vol.10 (4) 435-440 (1996)

[3] Luo Slgland et al, *J. Non-Crystalline Solids* No. 100 254-262 (1998)

[4] Runsheng Li ,*Basic Aluminium Chloride* p25 China Architecture & Building Press (1981)

[5] Xicheng Wang , Zhiqiang Cheng et al, *Chinese Materials Engineering* No.5 16-19 (1998)

[6] Shulan Shi, "Chemical Method for Preparing (Si-,Al-) Ceramic Wood" *Thesis for Master's Degree*

TABLE 1   Thermogravimetric parameters of woods

Sample	Range of thermo-decomposite temperature(^{0}C)	Weight loss(%) (at the end of thermo-decomposite)	weight loss (%)at 312^{0}C	Residual weight (%) at 745^{0}C
Untreated Wood	278-397	76.4	18.0	8.2
Composite A	312-407	59.7	14.2	16.9
Composite B	297-394	55.2	9.7	24.2

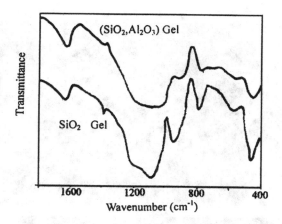

Fig.1 Infrared Spectra from Silica and (SiO_2 + Al_2O_3) gels

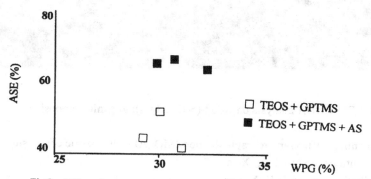

Fig.3 Effect of AS on the ASE of wood-inorganic composites

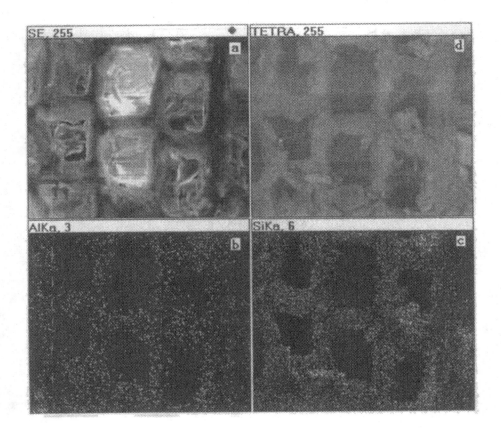

Fig.2 A SEM-EXDA analysis on wood-(SiO_2, Al_2O_3) inorganic composites

  a. A scanning electron micrograph of wood-(SiO_2, Al_2O_3) inorganic composites
  b. Distribution map of Al-K_α X-rays
  c. Distribution map of Si-K_α X-rays
  d. Distribution of the total components of wood-(SiO_2, Al_2O_3) inorganic composites

REAL-TIME MONITORING OF STRIATION DEVELOPMENT DURING SPIN-ON-GLASS DEPOSITION

Dylan E. Haas and Dunbar P. Birnie III
Department of Materials Science and Engineering
University of Arizona
Tucson AZ, 85721

ABSTRACT

A commercial spin-on-glass (SOG) solution was monitored during its deposition across silicon wafers in order to determine the moment at which coating defects, specifically striations, develop within the spinning solution. The radially oriented thick film ridges of striations act as a diffraction grating for an incident laser beam. Video monitoring of this diffraction pattern throughout the spin coating process allows us to determine the exact moment at which the striation defects begin to develop within the spinning solution. Simultaneous interferometric analysis has been performed to determine the fluid thickness at this instant in time. Performing this simultaneous video monitoring and interferometry during the spin coating of several Si wafers at various rotation rates we were able to confirm that striation growth is an evaporation driven phenomenon, taking place near the end of the "spin off" stage of spin coating.

INTRODUCTION

Spin coating is a common method for depositing thin, uniform films across planar substrates. Rapid wafer rotation causes the precursor fluid to spread radially and (theoretically) evenly across the surface. The coating fluid consists of solutes in solution with more volatile solvents that serve to lower the fluid viscosity and that steadily evaporate throughout the spin coating process leaving behind a thin, solute-rich film.

Spin coating is used in the microelectronics industry primarily for the deposition of photoresists in lithographic circuitry definition, although it has become increasingly important for the deposition of spin-on-glass (SOG) solutions. These SOGs are used primarily to provide planarization of substrates. The SOG solution can coat and fill in topographically complex surfaces that result after metallization of a particular level. The SOG also serves as a passivation and insulating dielectric layer. In addition, spin-on-glass solutions may be used to replace thermally grown gate and field effect oxide passivation layers in the manufacturing of various metal-oxide semiconductor (MOS) devices [1].

The spin coating process is traditionally broken into a number of stages [2,3] describing the physical mechanisms that dominate the thinning behavior over a period of

---

To the extent authorized under the laws of the United States of America, all copyright interests in this publication are the property of The American Ceramic Society. Any duplication, reproduction, or republication of this publication or any part thereof, without the express written consent of The American Ceramic Society or fee paid to the Copyright Clearance Center, is prohibited.

time. Essentially these can be summarized as the "spin-off," or "flow dominated," and "evaporation dominated" stages. "Spin-off" refers to the bulk removal due to centrifugal acceleration of fluid at the beginning. This stabilizes into a regime where the viscous effects control the thinning. This is followed immediately by the "evaporation" stage in which any further thinning occurs mainly due to the evaporation of solvents.

Although it is generally quite an efficient process, a number of coating defects can occur as a result of the spin coating process. One common type of surface defect is called "striations". These are thick ridges that develop during spinning and that emanate spoke-like from the center of the substrate. As feature sizes and layering depths decrease in IC and VLSI devices, the necessity for any SOG layers to be uniform and strictly planar increases. For these reasons it is imperative that spin on glass layers be as uniform and planar as possible. This requires intimate knowledge of the parameters governing striation growth during the spin coating process.

In the present work we are concerned with determining the precise moment at which striations begin to develop during the spin coating of a particular striation-prone SOG solution as well as the thickness of the film at this instant. Our overall objective is to attain a complete understanding of the factors governing spin-on processing of films.

EXPERIMENTAL PROCEDURE

In recent work [4] we have presented a simple diffraction experiment with which the characteristic spacing of the striations in a thin coating can be determined. In performing the present tests we have utilized the diffraction effect, but have applied it to wafers that are spinning rather than stationary; the motion of the wafer does not interfere with the striation diffraction effect. Thus, we have observed the light pattern from a laser reflecting from the surface of the wafer (with SOG solution flowing over it) throughout the spin coating process. This has allowed us to monitor when the striations develop. The independence of the diffraction effect between stationary and rotating wafers was confirmed by the fact that the diffraction pattern was not changed when the wafer stopped at the end of spinning. In earlier work we had shown that the average striation wavelength was dependent on the spinning rate [5], but it is also true that simultaneously the coating will be reaching a different final thickness. Independent of what the final coating thickness will be, the interesting thing to note is that for spin coating using a striation-producing solution, the monitoring of the reflected laser light reveals a distinct period during which the diffraction pattern develops – and by inference it indicates the time when the striation defects are formed and printed into the coating thickness.

The experimental setup is as follows: a 633 nm solid-state laser is aimed at a position roughly 1.5 in. from the center of the four-inch diameter Si wafers upon which the SOG is to be deposited. The laser is incident upon the wafer at a glancing angle of about 79° from normal. The reflected light rays then pass through a lens and the resulting diffraction pattern is imaged on a screen located at the focal length of the lens. A commercial Sony® digital video camera, model number DCR-TR7000, was then used to record the evolution of the diffracted light pattern throughout the spin coating process. The standard video frame rate of 30 fps then defined the time frequency at which this diffraction data was recorded. The moment at which the diffraction from striations was first noticeable was then defined as t_s and is tabulated below for several spin speeds.

At the same time, optical interferometry was performed upon the spinning solution. This optical interferometry follows along the precise scheme described elsewhere [6] for the measurement of evaporation rate and kinematic viscosity of spin-on solutions. A second laser is directed at near-normal incidence at a position near the center of the Si wafer. This light is passed through a beam splitter so that half is directed into a ("baseline") linear photo-detector and the other half travels straight down to reflect from the wafer/film surface. This reflected beam upon returning to the beam splitter is then directed into a second ("signal") linear photo-detector. The signal is then normalized using the baseline and the resulting intensity vs. time plot is analyzed. The reflected signal consists of a superposition of the light that is reflected from the fluid surface and that which is reflected from the surface of the Si wafer. These reflected signals are then translated into varying intensity vs. time plots, the peaks of which correspond to fluid thicknesses being an integral number of ¼ wavelengths (within the solution) thick. Interested readers are directed to [6,7] for a more thorough explanation as well as examples of these intensity vs. time plots, or "optospinograms [7]." With knowledge of the final coating thickness, we can use these "optospinograms" to generate a consistent plot of the fluid thickness H throughout the spin coating process. Using this H vs. t plot we can accurately determine the thickness of the fluid at the moment when striation growth gets started, H_s.

Using the processes described above, we monitored the spin coating of a commercially available SOG solution spun at different rotation rates varying from 500rpm to 4000rpm. Each trial began with the deposition of an approximately 5ml puddle of solution across the center of four-inch diameter Si wafers. This volume of fluid corresponds to roughly one millimeter of initial fluid thickness. The wafer was then accelerated at the highest ramping rate available, which brings the substrate from rest to the desired rotation rate within a half a second or less, as determined from independent stroboscopic video analyses performed upon the spin coater itself.

In order to determine the final coating thickness we etched away a small region of the SOG until all that remained was bare silicon and then measured the SOG thickness using a Tencor brand Alpha Step 200 surface profiling device.

RESULTS AND DISCUSSION

Table I and Fig.1 show the thickness H_s of the thinning film at the point when striations begin to develop and its dependence on the rotation rate ω. The exact relationship between H_s and ω is not straightforward. Indeed Fig.1 suggests that H_s takes on two distinct forms depending upon the range of rotation rates being considered. Fig. 1 also demonstrates a similar discontinuity in the amount of time t_s that passes before the thinning fluid attains this H_s depth at any particular spin speed.

The diffraction pattern imaged at a given spin speed reaches its full spread within a few seconds of the onset of striations, again depending upon the rotation rate. That is, once a diffraction pattern starts to become evident at t_s, it takes approximately 1 second for the responsible striation pattern to fully develop when spinning at 4000 rpm and 8 to 10 seconds at 500 rpm. As is depicted in Fig. 2 the fluid thickness H_s at the onset of striations is considerably less than the initial fluid depth. It is interesting to note that the solution depths are much smaller than the corresponding striation wavelengths (see last

Table I: Parameters characteristic of striation growth at several spin speeds, ω. H_s represent the fluid depth at the onset of striation growth; t_s is the time at which this growth begins, as referenced to the start of spin; Δt is the time period through which the striation pattern grows into its final form; H_f is the final thickness of the deposited film and λ_f is the final spacing between striations.

ω (rpm)	H_s (μm)	t_s (s)	Δt (s)	H_f (μm)	λ_f (μm)
500	8.0	4.1	9.6	1.5	143
1000	4.0	2.9	3.0	1.0	107
1500	2.6	2.4	1.9	0.80	89
2000	2.5	2.1	1.8	0.62	56
2500	4.2	1.1	1.8	0.56	44
3000	5.3	0.8	1.3	0.52	28
3500	5.7	0.7	1.3	0.46	29
4000	6.7	0.6	1.1	0.45	29

column in Table 1). In fact, the coating depths measured at the point when striations just get started is very small suggesting that our coating solutions have already reached a point where viscous outflow has effectively ceased and evaporation of the solvents controls the thinning of the solution.

We have demonstrated elsewhere [5] that the spacing between the ridges on a given striated wafer is constant across the entire substrate surface. Thus unlike the simple spoke analogy, the distance between these radially oriented thick film ridges remains essentially unchanged at any position on the film surface. That is, looking out along a given radius we see that striations appear to bifurcate locally once sufficient

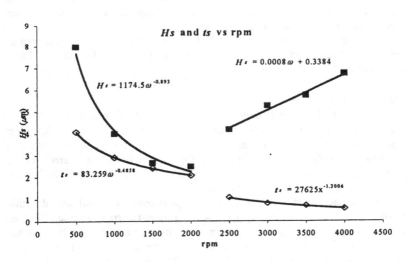

Fig.1: The variance in fluid thickness at the onset of striation growth H_s with spin speed ω as well as the variance in the amount of time which passes from the start of spin coating process to the onset of striation growth, t_s, with spin speed ω.

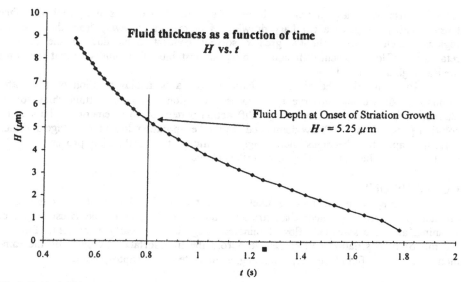

Fig 2: Typical thickness vs. time profile generated from interferometric monitoring of the thinning fluid behavior. This particular curve shows the thinning of an SOG spin coated at 3000 rpm. The vertical line represents the time at which striations began to develop as determined from frame by frame analysis of the video monitoring of the ensuing diffraction pattern which develops as the striations grow. The position at which this line crosses the H vs.t curve represents H_s, the fluid depth at the onset of striation growth.

spreading occurs such that the average spacing between striations remains constant across the entire substrate.

The development of striations as a direct result of solvent evaporation has been demonstrated experimentally [8,9]. The generation of ridges throughout the striation development time Δt which results in a constant striation spacing at all positions across the substrate can be explained in terms of localized feedback processes in solvent vapor evaporation from the SOG solution to the surrounding ambient. As the thinning solution nears the H_s fluid depth evaporation effects become increasingly significant. The evaporation rate becomes large enough that the solution's composition starts to be affected. A condition develops in which solvent diffuses faster out of particular regions near the fluid surface resulting in localized depletion of solvents and thereby localized surface tension variations (i.e. causing the "Marangoni Effect" [10]).

The airflow over these developing striations which is a result of the rotating substrate carries the diffused solvents tangentially "down stream" a brief distance prior to their ultimate bridging of the diffusion layer between the surrounding ambient and the fluid surface. Thus the air directly above the region behind a growing striation ridge is slightly more solvent enriched. This serves to decrease the diffusion of solvents from the solution in this region thereby enhancing the slight surface gradient in solvent concentration that fueled the striation growth in the first place. Furthermore the striation ridges themselves, being relatively depleted of solvents, possess a slightly higher surface tension which instigates a capillary drawing of more solvents away from these nearby regions which possess slightly greater solvent concentrations due to the air-flow retardation of local solvent diffusion, creating an unstable effect similar to that found for the Rayleigh instability [11].

By monitoring the diffraction pattern from a particular position on the wafer throughout Δt we can observe this process in action. Once the fluid thins to the characteristic H_s depth we see a slight diffraction pattern begin to emerge. The spin on solution passes from the "flow dominated" to the "evaporation dominated" stage and thus solvent evaporation becomes increasingly significant to the thinning process and the striation "seeds" begin to develop as described above.

CONCLUSIONS

Judging from the relative thickness of the SOG solution at the onset of a diffraction pattern it is clear that striations do not develop until the latest stages of spinning, after the so-called "flow dominated" stage is very nearly completed. Further, the full or final striation pattern sets in across the film surface at essentially the same moment the entire fluid layer stops thinning, as indicated by the optospinograms.

ACKNOWLEDGEMENTS

The work of our group has been facilitated greatly by an equipment loan from Specialty Coating Systems, Inc. And, the financial support of the National Science Foundation under grant DMR 98-02334 is very warmly appreciated.

REFERENCES

[1] C.W. Nam and S.I. Woo, "Characterization of Spin-Coated Silicate and Phosphosilicate Thin Films Prepared by the Sol-Gel Method," *Thin Solid Films*, **237** 314-19 (1994).
[2] P.C. Sukanek, "Spin Coating," *J. Imag. Tech.*, **11** [4] 184-90 (1985).
[3] D. Meyerhofer, "Characteristics of Resist Films Produced by Spinning," *J. Appl. Phys.* **49** [7] 3993-97 (1978).
[4] D.E. Haas and D.P. Birnie, III, "Diffraction Measurements of Striation Spacing for Spun-On Coatings," Manuscript in preparation.
[5] D.E. Haas, D.P. Birnie, III, M. J. Zecchino and J.T. Figueroa, "The Effect of Radial Position and Spin Speed on Striation Spacing in SOG Coatings," submitted to Thin Solid Films.
[6] D.P. Birnie, III and M. Manley, "Combined Flow and Evaporation of Fluid on a Spinning Disk," *Phys. Fluids*, **9** [4] 870-5 (1997).
[7] F. Horowitz, E. Yeatman, E. Dawnay and A. Fardad, "Real-Time Optical Monitoring of Spin Coating," *J. Phys. III France*, **3** 2059 (1993).
[8] X. M. Du, X. Orignac, and R. M. Almeida, "Striation-Free, Spin-Coated Sol-Gel Optical Films,"*J. Am. Ceram. Soc.*,**78** [8] 2254-56 (1995).
[9] B.K. Daniels, C.R. Szmanda, M.K. Templeton,& P. Trefonas III. "Surface Tension Effects in Microlithography-Striations," *Advances in Resist Technology and Processing III.*, **631** 192-201 (1986).
[10] L. E. Scriven and C. V. Sternling, "The Marangoni Effects", *Nature*, **187** 186-88 (1960).
[11] Lord Rayleigh, *Proc. London Math Soc.* **10**, 4 (1878).

KEYWORD AND AUTHOR INDEX

Alkoxides, 111
Antonietti, M., 49
Aparicio, M., 73
Applications, 15

$BaTiO_3$, 105
Baskaran, Suresh, 39
Biosensors, 1
Biotechnology, 1
Birnbaum, Jerome C., 39
Birnie, Dunbar P., III, 133

Cao, Guozhong, 87, 93
Catalysts, 127
Cell therapy, 1
Ceramic fibers, 111
Coatings, 121
Commercialization, 15, 29
Composites, 127
Condensation, 81
Coyle, Christopher A., 39

Dimethylsiloxane, 99
Drug delivery, 1
Dunham, Glen, 39

Ferroelectric films, 93
Flat-panel displays, 1
Fleig, P.F., 121
Flexible sheets, 99
Florian, P., 49
Forbess, M.J., 87, 93
Förster, S., 49
Friction, 121
Fryxell, Glen E., 39

GC spectra, 81
Glasses, 73
Göltner, C., 49

Haas, Dylan E., 133
Han, Sang-Mok, 81
Hou, Hongmei, 127
Hybrid materials, 49
Hydrolysis, 81

Inorganic/organic hybrids, 99
IR spectra, 81
Isota, Yoshiro, 105

Japanese industries, 29
Jewell-Larson, Nels, 87

Kang, Sang-Kyu, 81
Kang, Wi-Soo, 81
Katayama, Katsumi, 105
Katayama, Shingo, 99
Kawakami, Keiko, 99
Kim, Kyung-Nam, 81
Klein, L.C., 73
Kohler, Nathan, 39
Kozuka, Hiromitsu, 105

Lanthanides, 117
Li, Xiaohong, 39
Limmer, S.J., 87, 93
Liu, Jun, 39

Mechanical properties, 99
Metal alkoxides, 99

Nanoparticle technology, 15
Nanoporous dielectric films, 39
Nickel titanate, 121
NMR spectroscopy, 49

Page, R.A., 121
Patterned microstructures, 87
Piezoelectricity, 87, 111
Pope, Edward J.A., 1
PZT (lead zirconate titanate), 87, 105

$SrBaNb_2O_6$, 111
Sakka, Sumio, 29
Sanchez, C., 49
Schmidt, Helmut, 15
Seraji, Seana, 87, 93
Shi, Peng, 127
Shi, Shulan, 127
Shin, Dae-Yong, 81
Shirono, Kyougo, 111
Soft lithography, 87

Sol–gel processing, 117
Spin coating, 133
Spin-on-glass deposition, 133
Steunou, N., 49
Striations, 133
Strontium niobate, 87
Sun, Li, 127
Synthesis, 39

Takenaka, Shinsuke, 105
Taylor, D.J., 121
Thick films, 105
Thin films, 105

Titanium oxo-organic clusters, 49
Toyoda, Masahiro, 111
Tribology, 121
Tungsten bronze structures, 111

Wang, Xicheng, 127
Wear, 121
Wood, 127
Woodhead, James L., 117
Wu, Yun, 87, 93

Yamada, Noriko, 99